Arthur Gardiner Butler

The Butterflies of Malacca

Arthur Gardiner Butler

The Butterflies of Malacca

ISBN/EAN: 9783337331252

Printed in Europe, USA, Canada, Australia, Japan

Cover: Foto ©berggeist007 / pixelio.de

More available books at **www.hansebooks.com**

[533]

XXI. *The Butterflies of Malacca**. By ARTHUR GARDINER BUTLER, *F.L.S, F.Z.S.,* Senior Assistant in the Zoological Department, British Museum.

(Plates LXVIII., LXIX.)

Read December 21st, 1876.

THE generosity of Captain Stackhouse Pinwill, who has presented to the British Museum the whole of the Lepidoptera collected by himself during some years' residence in Malacca and Penang, has enabled me to study (with, I trust, some degree of profit to others as well as to myself) the Lepidopterous fauna of the Malayan peninsula. As I was obliged to go over the whole of the species hitherto recorded from Malaysia, and as Captain Pinwill's collection is the most complete one hitherto received from the peninsula, I determined not to lose the present opportunity of making a complete list of the Butterflies as yet known to occur at Malacca, comparing them at the same time, by means of a table, with the species hitherto received from India, China, Siam, and the islands to the south of the peninsula.

Of the 258 species now registered from Malacca, 36 appear to be endemic; of the remainder rather more than a fourth occur either at Assam or Nepal, more than a seventh at Moulmein, less than a seventh at Ceylon, nearly two fifths (apparently) in the island of Penang, about two elevenths at Singapore, about three sevenths in Borneo, about three sixteenths in Sumatra, more than a third in Java, about two thirteenths in Siam, rather more than a tenth in China, two species in the New Hebrides, and six in Australia. We see therefore that, with the exception of the last-mentioned eight species, the Butterflies of Malacca are limited to the Indian Region ; there are, however, several forms occurring in the Australian Region which some Lepidopterists would not regard as specifically distinct, such as the various allied forms of *Danais, Melanitis, Mycalesis, Doleschallia, Neptis, Diadema, Cynthia, Junonia, Lampides, Amblypodia, Calopsilia,* and *Papilio.* All of these are, nevertheless, characterized by slight but constant differences, and consequently have a right to be regarded as distinct species.

It appears to me to be important to those who desire conscientiously to study the geographical distribution of animals, to discriminate between even the most nearly allied species. I believe that few things have more retarded zoological geography than the reckless association together of so-called "local varieties" under the same specific name. I will cite an instance in illustration of what I mean. Some Lepidopterists still assert that *Diadema bolina* ranges from Northern India to New Zealand, whereas that species does not occur outside the Indian Region. Allied forms, indeed, are common throughout the Moluccas, Australia, and the South Pacific; and one of them (*D. nerina*), strange

* Two of the new species mentioned in the present paper, and an abstract of the whole, have already appeared in the Society's Journal (Zool. vol. xiii. p. 115 and p. 196 respectively).

enough, is found both in Java and the Australian Region, whilst a dwarfed form of it is taken in the Philippines; yet *D. nerina*, although more widely distributed than *D. bolina*, exhibits even less individual variation.

The fact of two species of *Terias*, discovered in the New Hebrides, turning up in Malacca, is somewhat startling; but it is quite possible, in a genus where only a practised eye can at once detect the specific differences, that the same form of imago may have been simultaneously developed in both localities; moreover, breeding might (as it often does) reveal constant and well-marked distinctive characters whereby the earlier stages could at once be separated.

The following Note, which appears to me to be worthy of the consideration of naturalists, has been communicated by my friend Mr. W. L. Distant, who was for some years resident at Province Wellesley :—

" All catalogues of the Penang fauna must be accepted with some qualifications. The dependencies of Penang and Province Wellesley are usually made more or less into one zoological province, especially as regards the insects. This I found to be the case when collecting there a few years ago, all collections being then made indiscriminately from both localities.

" The island of Penang is about 20 miles long by 9 miles broad, comprising an area of about 107 square miles. A group of granite hills occupies about two thirds of its extent, running through its centre from north to south, bounded by a plain on their eastern and western sides. These hills are densely covered with a beautiful forest on all sides; and their highest point, West Hill, is about 2600 feet above the sea.

" Province Wellesley is situated on the Malayan peninsula, and is separated from Penang by a strait from 2 to 10 miles broad; it is about 35 miles in length, averaging 8 miles in width. Although it possesses many hills of a considerable elevation (yet not equalling those of the island of Penang), its general character is more of a plain, in which jungle tracts and cultivated lands of rice and sugar-cane are its characteristics. Mangrove swamps, often of great extent, exist in the neighbourhoods of its many creeks and rivers.

" There is little doubt that, when the floras and faunas of these two districts are worked out separately and distinctly, a common character will be found to pervade them both, but that many species will be found locally distinct and constant. Species may be expected to exist in the lofty wooded regions of Penang that are absent in the plains and less elevated hills of Province Wellesley. At present, however, when mixed collections are made from both regions, the habitat ' Penang' must be accepted as not representing the insular fauna alone, but comprising also that of a portion of the Malayan peninsula."

List of Species collected by Captain Pinwill.

RHOPALOCERA.

Family NYMPHALIDÆ.

Subfamily DANAINÆ, Bates.

Genus EUPLŒA, Fabricius.

1. EUPLŒA PHŒBUS, Butler, Proc. Zool. Soc. p. 270. n. 3 (1866).   ♂ ♀, Malacca. This is clearly the species described by Felder as *E. Castelnaui*.

2. EUPLŒA OCHSENHEIMERI, Lucas, Rev. et Mag. Zool. p. 315 (1853).   ♂ ♀, Malacca.

3. EUPLŒA GROTEI, Felder, Reise der Nov., Lep. ii. p. 339. n. 470, pl. 41. fig. 7 (1865).   ♂ ♀, Malacca.

4. EUPLŒA MARGARITA, Butler, Proc. Zool. Soc. p. 279. n. 34 (1866).   ♂ ♀, Malacca; ♀, Penang.

5. EUPLŒA CHLOË, Guérin, in Deless. Souv. Voy. d. l'Inde, ii. p. 71 (1843).   ♂ ♀, Malacca.

6. EUPLŒA BREMERI, Felder, Wien. ent. Monatschr. iv. p. 398. n. 16 (1860).   ♂, Malacca. Mr. W. L. Distant took this species at Province Wellesley; he says that all the species of *Euplœa* which he captured were flying in the months of August and September.

7. EUPLŒA MÉNÉTRIÉSI, Felder, Wien. ent. Monatschr. iv. p. 398. n. 15 (1860). Malacca.

8. EUPLŒA PINWILLI, n. sp. (Pl. LXIX. fig. 9.) Primaries brownish piceous, purple-shot, slightly paler along the external border: a long sericeous interno-median streak : secondaries paler brown, deepest at base; two marginal series of whitish spots, clear at anal angle, obsolescent and decreasing towards apex; costal area greyish. Primaries below paler than above, with a subcostal spot and an elliptical interno-median spot pinky white; a spot in the cell and two beyond it bluish white, two or three white dots at external angle; secondaries with the submarginal spots clear whitish; a spot in the cell and seven in an arched series beyond it lilacine whitish; base and pectus white-spotted. Expanse of wings 4 inches. ♂ ♀, Malacca. *E. Pinwilli* is allied to *E. Ménétriési*, but larger, darker, and shot with purple.

9. EUPLŒA MIDAMUS, (*Papilio midamus*) Linnæus, Mus. Lud. Ulr. p. 251 (1764). ♂ ♀, Malacca; ♀, Penang. "This species is very common in gardens" (*W. L. Distant*).

Genus CALLIPLŒA, Butler.

1. CALLIPLŒA DIOCLETIANUS, (*Papilio diocletianus*) Fabricius, Ent. Syst. iii. 1, p. 40. n. 118 (1793).   ♂ ♀, Malacca. "Common in gardens" (*W. L. Distant*).

2. CALLIPLŒA VESTIGIATA, (*Euplœa vestigiata*) Butler, Proc. Zool. Soc. p. 288. n. 58, fig. 1 (1866).   ♂ ♀, Malacca.

3. CALLIPLŒA LEDERERI, (*Euplœa Ledereri*) Felder, Wien. ent. Monatschr. iv. p. 397. n. 14 (1860); Reise der Nov., Lep. ii. p. 317. n. 431, pl. 40. figs. 5, 6 (1865). ♂ ♀, Malacca.

Genus SALPINX.

1. SALPINX LEUCOGONIS, n. sp. Nearly allied to *E. vestigiata*, but smaller, the costa of primaries not so strongly arched, the outer margin slightly inarched in the centre, the submarginal spot on second median spot wanting, the subcostal spot above the end of the cell much smaller, the spot on interno-median area widened into a notched blotch, all the spots lilac; secondaries paler, only three lilacine whitish spots placed obliquely near apex, no submarginal spots, anal angle white. Primaries below with only two or three submarginal and three or four marginal white dots; secondaries with no white blotches beyond the cell. Expanse of wings 3 inches 2 lines.   ♀, Malacca.

Genus DANAIS, Latreille.

1. DANAIS PLEXIPPUS, (*Papilio plexippus*) Linnæus, Mus. Lud. Ulr. p. 262 (1764). ♂ ♀, Malacca; ♂, Penang. " Common in August and September " (*W. L. Distant*).

2. DANAIS MELANIPPUS, (*Papilio melanippus*) Cramer, Pap. Exot. ii. pl. 127. figs. A, B (1779).   ♂ ♀, Malacca; ♂, Penang.

3. DANAIS VULGARIS, Butler, Ent. Mo. Mag. p. 164 (1874).   ♂ ♀, Malacca.

4. DANAIS SEPTENTRIONIS, Butler, Ent. Mo. Mag. p. 163 (1874).   ♂, Malacca; ♀, Penang.

5. DANAIS GRAMMICA, Boisduval, Sp. Gén. Lép. i. pl. 11. fig. 10 (1836).   ♂ ♀, Malacca.

6. DANAIS MELANEUS, (*Papilio melaneus*) Cramer, Pap. Exot. i. pl. 30. fig. D (1775). ♀, Malacca.

7. DANAIS CROCEA, Butler, Proc. Zool. Soc. p. 57. n. 53, pl. 4. figs. 5, 6 (1866). ♂ ♀, Malacca. " Occurs at Penang in August and September " (*W. L. Distant*).

Genus IDEOPSIS, Horsfield.

1. IDEOPSIS DAOS, (*Idea daos*) Boisduval, Sp. Gén. Lép. i. pl. 24. fig. 3 (1836). ♂ ♀, Malacca; ♀, Penang.

Genus HESTIA, Hübner.

1. HESTIA LYNCEUS, (*Papilio lynceus*) Drury, Ill. Ex. Ent. ii. pl. 7. fig. 1 (1773). ♂, Malacca. " The species of *Hestia* go by the local name of ' the widow ' in Province Wellesley; the genus was very abundant near the top of Penang Hill " (*W. L. Distant*).

2. HESTIA LINTEATA, n. sp. (Pl. LXIX. fig. 6.) Nearly allied to *H. belia*, but much larger, the veins less broadly black-bordered; spots larger, excepting towards the costa of primaries at apex; discoidal spot of secondaries very large; clavate markings terminating the veins much longer, more slender in the middle. Expanse of wings 5 inches 6 lines to 6 inches 8 lines.   ♂ ♀, Malacca (six examples).

Subfamily SATYRINÆ, Bates.

Genus MELANITIS, Fabricius.

1. MELANITIS LEDA, (*Papilio leda*) Linnæus, Syst. Nat. i. 2, p. 773. n. 150 (1766). ♂ ♀, Malacca, Penang. "Common in jungle-grass " (*W. L. Distant*).

Genus LETHE, Hübner.

1. LETHE EUROPA, (*Papilio europa*) Fabricius, Syst. Ent. p. 500 (1775). ♂, ♀, Malacca; ♂, Penang. " In grassy places" (*W. L. Distant*).

Genus MYCALESIS, Hübner.

1. MYCALESIS MINEUS, (*Papilio mineus*) Linnæus, Syst. Nat. i. 2, p. 768. n. 126 (1766). ♂ ♀, Malacca; ♀, Penang.

2. MYCALESIS POLYDECTA, (*Papilio polydecta*) Cramer, Pap. Exot. ii. pl. 144. figs. E, F (1779). ♂, Malacca, Penang.

3. MYCALESIS JANARDANA, F. Moore, Cat. Lep. E.I. C. i. p. 234 (1857). ♂, Malacca.

4. MYCALESIS HESIONE, (*Papilio hesione*) Cramer, Pap. Exot. i. pl. 11. figs. C, D (1779). ♂ ♀, Malacca, Penang.

5. MYCALESIS FUSCA, (*Dasyomma fuscum*) Felder, Wien. ent. Monatschr. iv. p. 401. n. 27 (1860). Malacca.

YPTHIMA, Westwood.

1. YPTHIMA PHILOMELA, Hübner, Zutr. ex. Schmett. figs. 83, 84 (1806). ♂, ♀, Malacca; ♀, Penang.

2. YPTHIMA (*sic*) METHORA, Hewitson, Trans. Ent. Soc. 3rd ser. ii. p. 291, (*Ypthima*) pl. 18. figs, 20, 21 (1865). ♂ ♀, Malacca; ♀, Penang.

3. YPTHIMA CORTICARIA, n. sp. Wings above greyish brown, an inner submarginal streak and an outer submarginal line blackish diffused; primaries with a large subapical oval black ocellus, with two plumbaginous pupils and a testaceous iris surrounded by a dusky line; a similar smaller unipupillated rounded ocellus near anal angle of second-aries. Wings below white, densely reticulated with brown, two ill-defined, interrupted, central, subparallel lines; a submarginal streak and line as above: primaries with the ocellus brighter and better defined than above; outer border pale brown, bounded by the inner submarginal streak: secondaries with three ocelli, one at apex (below which is sometimes another extremely small ocellus) and two placed obliquely at anal angle, the lower one small and irregular. Expanse of wings 1 inch 8 lines. Allied to *Y. nareda*. ♂ ♀, Malacca.

Subfamily ELYMNIINÆ, Kirby.

Genus ELYMNIAS, Hübner.

1. ELYMNIAS NIGRESCENS, Butler, P. Z. S. p. 520, pl. 42. fig. 1 (1871). ♂ ♀, Malacca, Penang. A long series of this species, exhibiting no variation.

2. ELYMNIAS MEHIDA, (*Melanitis mehida*) Hewitson, Exot. Butt. iii. *Mel.* pl. 1. figs. 2, 3 (1863). Malacca.

Subfamily MORPHINÆ, Butler.

Genus AMATHUSIA, Fabricius.

1. AMATHUSIA PHIDIPPUS, (*Papilio phidippus*) Linnæus, Syst. Nat. i. 2, p. 752 (1767). ♂ ♀, Malacca.

Two forms (probably seasonal varieties) occur in Malacca, the one with dark bands below, the other with pale bands; in the female of the dark form the subapical yellowish patch on the upper surface of primaries is more distinct. I have seen three examples in Mr. Distant's collection from Penang. It occurs in August and September.

Genus ZEUXIDIA, Hübner.

1. ZEUXIDIA AMETHYSTUS, Butler, P. Z. S. p. 485. n. 5 (1865). ♂ ♀, Malacca. The female differs from the male much in the same way as in the sexes of *Z. doubledayi*.

Genus DISCOPHORA, Boisduval.

1. DISCOPHORA MENETHO, (*Papilio menetho*) Fabricius, Ent. Syst. iii. 1, p. 83. n. 260 (1793). ♂, Malacca.

2. DISCOPHORA TULLIA, (*Papilio tullia*) Cramer, Pap. Exot. i. pl. 81. figs, A, B (1779). ♂ ♀, Malacca. The typical male differs from that of the Indian form in the unspotted character of the upper surface.

Genus XANTHOTÆNIA, Westwood.

1. XANTHOTÆNIA BUSIRIS, (*Clerome busiris*) Westwood, Trans. Ent. Soc. ser. ii. vol. iv. p. 187 (1858). ♀, Malacca.

Genus THAUMANTIS, Hübner.

1. THAUMANTIS NOUREDDIN, Westwood, Gen. Diurn. Lep. p. 337. n. 6, note (1851); Trans. Ent. Soc. ser. ii. vol. iv. p. 175. pl. 20 (1858). ♀, Malacca. "Occurs at Penang in August and September" (*W. L. Distant*).

2. THAUMANTIS PSEUDALIRIS. (Pl. LXVIII. fig. 1.) *Thaumantis aliris* ♂, Westwood, Trans. Ent. Soc. n. s. vol. iii. p. 176 (1856-58); Butler, Journ. Linn. Soc. xiii. p. 115. This species differs from the male of *T. aliris* from Borneo in having the band of primaries above half the width, not notched, and yellower in tint; the basal area of all the wings ferruginous. Below, the area beyond the band is not striated, and is of the same rusty red colour as the broad outer border, the latter is also paler at the edge, and is cut much shorter by the obliquity of the transverse band; the basal spots are rusty red instead of red-brown and black; the irregular band of secondaries is paler and much more constricted in the centre; the external and discal areas are altogether paler and more uniform in tint; the ocelli are considerably smaller and paler; the spots between

the occlli and the yellow submarginal spots are obsolctc. Expansc of wings 4 inchcs 7 lincs. ♂, Malacca.

As Professor Westwood figures the Bornean female, I am obliged to rcnamc the Malacca male.

Subfamily NYMPHALINÆ, Bates.

Genus DOLESCHALLIA, Felder.

1. DOLESCHALLIA PRATIPA, Felder, Wicn. ent. Monatschr. iv. p. 399 (1860). ♂ ♀, Malacca. This spccics has hitherto bcen represented in the Museum collection by only one female; it differs considerably from *D. bisaltide* in both sexes.

Genus CHARAXES, Ochsenhcimer.

1. CHARAXES JALYSUS, Felder, Rcisc der Nov., Lep. iii. p. 438, pl. 59. fig. 5 (1867). ♂, Malacca. Mr. Distant gave me an cxamplc of this species taken by him at Provincc Wcllcsley.

2. CHARAXES HEBE, Butler, P. Z. S. p. 634, pl. 37. fig. 3 (1865). ♂ ♀, Malacca. The male does not differ at all in pattern from the female.

3. CHARAXES SCHREIBERI, (*Nymphalis Schreiber*) Godart, Enc. Méth. ix. Suppl. p. 825 (1823). ♂, Malacca.

Genus ADOLIAS, Boisduval.

1. ADOLIAS ADONIA, (*Papilio adonia*) Cramer, Pap. Exot. iii. pl. 255, figs. C, D (1782). ♂ ♀, Malacca.

2. ADOLIAS GARUDA, Moore, Cat. Lep. E.I. C. i. p. 186. n. 374 (1857). ♂, Malacca.

3. ADOLIAS JAMA, Felder, Rcisc der Nov., Lep. iii. p. 431 (1867). ♂ ♀, Malacca. This spccics appears to bc common.

4. ♀ ADOLIAS LAVERNA, Butler, Cist. Ent. i. p. 29 (1870). ♂ ♀, Malacca. Now that both sexes of this species have come from the same locality, I find that the male from Borneo figured in my 'Lepidoptera exotica' is a distinct species, the sccondaries of typical *A. laverna* being totally unrelicved by bright colouring, and much resembling those of *A. garuda*.

5. ADOLIAS ASOKA, Felder, Rcisc der Nov., Lep. iii. p. 433, pl. 58. fig. 1 (1867). ♂ ♀, Malacca. This spccics was not previously in the collection of the British Museum.

6. ADOLIAS MONINA, (♂ *Papilio monina*) Fabricius, Mant. Ins. ii. p. 51 (1787). ♂ ♀, Malacca. The female, although much like *A. puseda*, is rather different. The following is probably only a variation of *A. monina*.

7. ADOLIAS LUDEKINGII, Vollenhoven, Tijd. Ent. v. p. 189, pl. 10. fig. 3 (1860). ♂ ♀, Malacca.

Genus SYMPHÆDRA, Hübner.

1. SYMPHÆDRA DIRTEA , (*Papilio dirtea*) Fabricius, Ent. Syst. iii. 1, p. 59. n. 184 (1793). ♂ ♀, Malacca, ♂, Penang.

" Slices of cut pineapple placed along a road that ran by this jungle, were generally sure in a short time, at the proper season, to be visited by a good supply of both males and females." "Old and fallen fruits of most kinds were attractive; but sliced pineapple was mostly used as bait" (*W. L. Distant* Ent. Mo. Mag. vol. xii. p. 207).

Genus TANAËCIA, Butler.

1. TANAËCIA ARUNA, (*Adolias aruna*) Felder, Wien. ent. Monatschr. iv. p. 400. n. 24 (1860). ♂, Malacca. Not previously in the Museum collection.

2. TANAËCIA PULASARA, (*Adolias pulasara*) Moore, Cat. Lep. E.I. C. i. p. 190 (1857). ♂ ♀, Malacca. The male is very closely allied to *T. varuna* of Vollenhoven, but has the apical area of secondaries white, and the band between the zigzag lines of primaries broader.

Genus ATHYMA, Westwood.

1. ATHYMA LARYMNA, Westwood & Hewitson, Gen. Diurn. Lep. pl. 35. fig. 1 (1850). ♂, Malacca.

2. ATHYMA LEUCOTHOË, (*Papilio leucothoë*) Linnæus, Mus. Lud. Ulr. p. 292 (1764). ♂, Penang.

3. ATHYMA KRESNA, Moore, P. Z. S. p. 12. n. 6, pl. 50. fig. 4 (1858).   ♂ ♀, Malacca.

4. ATHYMA SABRATA, Moore, P. Z. S. p. 13. n. 10, pl. 51. fig. 1 (1858).   ♂ ♀, Malacca.

5. ATHYMA NIVIFERA, n. sp.   (Pl. LXIX. fig. 4.)   *Athyma nefte*, Moore (ex parte), P. Z. S. 1858, p. 13.   This species may be at once distinguished from *A. nefte* of Cramer by the much narrower central band, which in the male is much more distinctly blue at the edges; by the longer and narrower trifid subapical band on the male, and the deeper colour of the underside. Expanse of wings 2 inches 5–9 lines.   ♂ ♀, Malacca. It is also common in Borneo, and occurs at Assam.

6. ATHYMA AMHARA, Druce, P. Z. S. p. 344, pl. xxxii. fig. 2 (1873).   ♂, Malacca. In the examples from Malacca the discoidal white streak of primaries is reduced to one or two white dots, connected by a pale brown streak.

7. ATHYMA CLERICA, n. sp. (Pl. LXIX. fig. 5.)   Nearly allied to *A. abiasa*, but larger, the spots of primaries more oblique and larger, the subbasal transverse white band of secondaries narrower; the discal series of spots much larger, more inarched above anal angle, forming a waved band, divided by the nervures; a well-marked greenish-grey submarginal streak.   Below much paler, with the differences of the upper surface; submarginal series of lituræ replaced by the submarginal streak, which is rosy greyish (not greenish, as above). Expanse of wings 2 inches 7 lines.   ♂, Malacca.

8. ATHYMA IDITA, Moore, P. Z. S. p. 16. n. 16, pl. 51. fig. 3 (1858). ♂ ♀, Malacca.

9. ATHYMA PRAVARA, Moore, Cat. Lep. E.I. C. i. p. 173, pl. 5a. fig. 4 (1857). ♂, Malacca.

Genus EURIPUS, Westwood.

1. EURIPUS EUPLŒOIDES, Felder, Reise der Nov. Lep. iii. p. 415. n. 638 (1867). ♂ ♀, Malacca. This species is a mimic of *Calliplœa diocletianus*.

Genus EULACURA, Butler.

1. EULACURA OSTERIA, (*Apatura osteria*) Westwood, Gen. Diurn. Lep. p. 305, note (1850). ♂ ♀, Malacca.

Genus LEBADEA, Felder.

1. LEBADEA ALANKARA, (*Aconthea alankara*) Horsfield, Cat. Lep. E.I. C. pl. 5. fig. 6 (1829). ♂, Malacca. This appears to be the male of the succeeding species.

2. LEBADEA MARTHA, (*Papilio martha*) Fabricius, Mant. Ins. ii. p. 56. n. 555 (1787). ♀, Malacca.

Genus LIMENITIS, Fabricius.

1. LIMENITIS PROCRIS, (*Papilio procris*) Cramer, ii. pl. 106. figs. E, F (1779). ♂ ♀, Malacca; ♂, Penang. "Very common at Province Wellesley" (*W. L. Distant*).

Genus NEPTIS, Fabricius.

1. NEPTIS MAMAJA, n. sp. (Pl. LXIX. fig. 3.) Nearly allied to *N. eurynome*, but always to be distinguished by the narrower externo-discal band of white spots in secondaries, colouring below rather brighter than in *N. eurynome*. Expanse of wings 1 inch 11 lines to 2 inches 4 lines. ♂ ♀, Malacca and Penang. This appears to be the commonest *Neptis* in Malacca.

2. NEPTIS LEUCONATA, n. sp. (Pl. LXIX. fig. 1.) Wings above olive-brown, banded and spotted with cream-colour; arrangements of markings as in *N. nata*, but the bands of secondaries wider and nearer together; pale brown intermediate streaks better defined. Wings below slightly redder; bands and spots as above; intermediate streaks white. Expanse of wings 2 inches. ♂, Malacca.

3. NEPTIS GONONATA, n. sp. (Pl. LXIX. fig. 2.) Wings of the same shape and with the same spots and bands as *N. soma*; but all these markings pure white instead of sordid yellowish white. Wings below chocolate-brown; bands below straighter; submarginal line of secondaries below white. Expanse of wings 2 inches 1 line. ♂, Malacca.   Intermediate in character between *N. nata* and *N. soma*.

4. NEPTIS NATA, Moore, Cat. Lep. E.I. C. i. p. 168, pl. 4a. fig. 6 (1857). ♂ ♀, Malacca.

Vikasi Group.

5. NEPTIS VIKASI, Horsfield, Cat. Lep. E.I. C. p. 168, pl. 5. figs. 2, 2a (1829). ♂, Malacca.

Columella Group.

6. NEPTIS COLUMELLA, (*Papilio columella*) Cramer, Pap. Exot. iv. pl. 296. figs. A, B (1782).   ♂ ♀, Malacca.

Tiga Group.

7. NEPTIS DORELIA, n. sp.   (Pl. LXVIII. fig. 3.)   *Papilio heliodore,* Fabricius (nec Cramer nec Moore), Mant. Ins. ii. p. 52. n. 516 (1787).   ♂ ♀, Malacca.

The type of the Fabrician species differs from the examples of *N. tiga* in the Horsfield Cabinet from Java in having a single instead of a double central arched line across the under surface of the secondaries; this line is, in some examples, slightly thickened.

8. NEPTIS TIGA, Moore, Proc. Zool. Soc. p. 4. n. 3 (1858).   ♂ ♀, Malacca. This species agrees with an example in the Banksian cabinet, but not with the Fabrician type; there are two central brown curved lines across the underside of the secondaries, partially or wholly filled in with paler brown.

9. NEPTIS PARAKA, n. sp.   (Pl. LXVIII. fig. 2.)   Almost exactly like *N. hordonia* above, but brighter in colour, and with the tawny bands rather broader; the submarginal tawny streak of primaries deeply bisinuate so as to form two wide arches, and intersected by a black line.   Wings below ochre-yellow, the black portions of the upper surface represented by brownish stains, and partially blotched with grey: the basal area of secondaries crossed by two black oblique lituræ; the centre band bordered by two inter-rupted black lines; the outer border intersected (above the indication of the submarginal tawny line of the upper surface) by a slightly waved black line.   Expanse of wings 1 inch 7 to 10 lines.   ♂ ♀, Malacca.   The typical *N. heliodora* very nearly resembles this species on the upper surface; but below it is very different.

10. NEPTIS DINDINGA, n. sp.   (Pl. LXVIII. fig. 6.)   Larger than *N. heliodora,* the bands broader, particularly the central band of secondaries, which is twice as broad and extends nearly to the base; subapical patch of primaries twice as wide, and deeply indented in the centre of its inner border; submarginal tawny streak replaced by pale brown (but intersected by the black line) in the primaries, tawny but extremely slender in the secondaries.   Differences below much as above; discal band of secondaries broad and black, and intersected by a whity-brown streak.   Expanse of wings 2 inches. ♀, Malacca.

We have a specimen in the British Museum from Moulmein, which measures 2 inches 2 lines in expanse.   It is a well-marked species.

11. NEPTIS HORDONIA, (♂, *Papilio hordonia*) Stoll, Suppl. Cram. pl. 33. figs. 4, 4 D (1790).   ♀, Malacca.

Genus PANDITA, Moore.

1. PANDITA SINOPE, Moore, Cat. Lep. E.I. C. i. p. 182, pl. 6a. fig. 3 (1857).   ♂, Malacca; Penang.

Genus HYPOLIMNAS, Hübner.

1. HYPOLIMNAS BOLINA, (*Papilio bolina*) Linnæus, Mus. Lud. Ulr. p. 295 (1764). ♂ ♀, Penang.

2. HYPOLIMNAS INCOMMODA, n. sp. ♂ very similar to the male of *H. bolina*, but with the subapical band of primaries straighter on the underside, and the pale brown submarginal spots narrower and darker: expanse of wings 3 inches 1 line. ♀ differs from the female of *H. bolina* in having a broad oblique subapical white band on the upperside of primaries, a large diffused sordid white patch just beyond the cell of secondaries, and the submarginal spots all separated, small, and pale brown: expanse of wings 3 inches 11 lines. ♂ ♀, Malacca.

This form seems to replace *H. bolina* in Malacca; in some respects it approaches the Javan species *H. nerina*.

Genus CETHOSIA, Fabricius.

1. CETHOSIA METHYPSEA, n. sp. Allied to *C. hypsina* and *C. penthesilea*: from the former it differs in its smaller expanse, the regularity of the subapical oblique whitish band of primaries, in the presence of an apical submarginal series of four whitish spots connecting the lower extremity of this band with the costa, and in the much less prominent angulation of the outer edge of the red band on the under surface of secondaries; from the latter it differs in its smaller expanse, much brighter coloration, the yellower tint of the subapical band of primaries, the more irregular and much more broadly white-bordered red band below; the primaries are also more produced. Expanse of wings 2 inches 9 lines to 3 inches 3 lines. ♂, Penang. The larger measurement is taken from examples previously in the collection. We have a female from Assam.

2. CETHOSIA HYPSINA, Felder, Reise der Nov., Lep. iii. p. 385 (1867). ♂ ♀, Malacca. Also in the Museum collection from Nepal and Assam.

Genus CIRROCHROA, Doubleday.

1. CIRROCHROA ORISSA, Felder, Wien. ent. Monatschr. iv. p. 399 (1863); Reise der Nov., Lep. iii. pl. 49. figs. 7. 8 (1867). ♂ ♀, Malacca; Penang.

2. CIRROCHROA JOHANNES, Butler, P. Z. S. p. 221, pl. 17. fig. 10 (1868). ♂ ♀, Malacca.

3. CIRROCHROA BAJADETA, Moore, Cat. Lep. E.I. C. i. p. 150, pl. 3a. fig. 3 (1857). ♀, Malacca. This seems to be Felder's *C. malaya*.

4. CIRROCHROA ROTUNDATA, n. sp. ♀. Nearly allied to *C. mithila* of Moore, the wings more rounded, primaries less produced; markings of primaries above almost obsolete; outer undulated line of secondaries much more distinct; below all the bands and spots tawny instead of ochre yellow: expanse of wings 2 inches 11 lines. Malacca. I have no doubt that this species is distinct from *C. mithila*; the coloration of the bands below is strikingly different.

4 c 2

Genus Terinos, Boisduval.

1. Terinos teuthras, Hewitson, P. Z. S. p. 80 (1862).   ♀, Malacca.

2. Terinos robertsia, Butler, Ann. & Mag. Nat. Hist. ser. iii. vol. xx. p. 400, pl. 8. figs. 2–4 (1867).   ♂ ♀, Malacca.   This species appears to be tolerably common.   Mr. Distant took both the above butterflies at Province Wellesley.

Genus Messaras, Doubleday.

1. Messaras erymanthis, (*Papilio erymanthis*) Drury, Ill. Ex. Ent. i. pl. 15. figs. 3, 4 (1773).   ♂ ♀, Malacca; Penang.

Genus Atella, Doubleday.

1. Atella phalanta, (*Papilio phalanta*) Drury, Ill. Ex. Ent. i. pl. 21. figs. 1, 2 (1773). ♂ ♀, Malacca; Penang.

2. Atella sinha, (*Terinos sinha*) Kollar, Hügel's Kaschm. iv. 2, p. 438 (1848). ♂, Malacca.

Genus Cynthia, Fabricius.

1. Cynthia erotella, n. sp.   Allied to *C. erota*, but much smaller, brighter in colour; the primaries generally comparatively shorter, the costal margin more arched, the ocelli smaller in the male, and the discal lunulated line much less strongly defined; secondaries of the female with four ocelli, with broad bright orange irides.   Wings below slightly more reddish in colouring, with scarcely a trace of pearly grey clouding on the external area.   Expanse of wings, ♂ 3 inches 2–8 lines, ♀ 3 inches 7 lines.   ♂ ♀, Malacca.

I have examined a good series of examples, and am therefore confident of the distinctness of this species.   In general size and the indistinctness of the lunulated discal line it agrees with *C. deione*; but it is altogether brighter in colour, with longer tails to secondaries, and a different arrangement of the lines on the under surface of the wings.   Mr. Distant took it at Province Wellesley.

Genus Parthenos, Hübner.

Parthenos lilacinus, n. sp. (*Minetra gambrisius*, Hewitson, Gen. Diurn. Lep. pl. li. fig. 2).   Differs from *P. gambrisius* in the lilac internal area of primaries and basal area of secondaries above; in the blue-green basal areas of secondaries below, and the much better defined series of black lituræ beyond the hyaline band of primaries.   Expanse of wings 4 inches 2 lines.   ♂ ♀, Malacca; Penang.

I have seen long series of both the Indian and Malayan forms, and am therefore satisfied that the differences are constant.   The Fabrician type agrees with our examples from Silhet.

Genus Cyrestis, Boisduval.

1. Cyrestis rahria, Moore, Cat. Lep. E.I. C. i. pl. 3a. fig. 2 (1857).   ♂ ♀, Malacca; ♀, Penang.

2. CYRESTIS NIVEA, (*Amathusia nivea*) Zinken-Sommer, Nova Acta. Acad. Cur. xvi. p. 138, pl. 14. fig. 1 (1831).   ♂, Malacca.   "Not scarce at Province Wellesley " (*W. L. Distant*).

Genus ERGOLIS, Boisduval.

1. ERGOLIS ARIADNE, (*Papilio ariadne*) Linnæus, Syst. Nat. i. 2, p. 778. n. 170 (1767). ♂ ♀, Malacca; ♀, Penang.

2. ERGOLIS MERIONE, (*Papilio merione*) Cramer, Pap. Exot. ii. pl. 144. figs. G, H (1779). ♂, Malacca; Penang.

Genus JUNONIA, Hübner.

1. JUNONIA IPHITA, (*Papilio iphita*) Cramer, Pap. Exot. iii. pl. 209. figs. C, D (1782). ♂ ♀, Malacca; ♀, Penang.

2. JUNONIA LAOMEDIA, (*Papilio laomedia*) Linnæus, Syst. Nat. i. 2, p. 772. n. 145 (1767).   ♂ ♀, Malacca.   "Common at Province Wellesley " (*W. L. Distant*).

3. JUNONIA LEMONIAS, (*Papilio lemonias*) Linnæus, Mus. Lud. Ulr. p. 277 (1764). ♂, Malacca; ♀ ♂, Penang.

4. JUNONIA ORITHYA, (*Papilio orithya*) Linnæus, Mus. Lud. Ulr. p. 278 (1764). ♂ ♀, Malacca.   "Common at Province Wellesley " (*W. L. Distant*).

5. JUNONIA ASTERIE, (*Papilio asterie*) Linnæus, Syst. Nat. i. 2, p. 769. n. 133 (1767). ♂ ♀, Malacca; Penang.

Genus RHINOPALPA, Felder.

1. RHINOPALPA FULVA, Felder, Wien. ent. Monatschr. iv. p. 399. n. 21 (1860). ♂, Malacca.

Family ERYCINIDÆ.

Subfamily NEMEOBIINÆ, Bates.

Genus ZEMEROS, Boisduval.

1. ZEMEROS EMESOIDES, Felder, Wien. ent. Monatschr. iv. p. 396. n. 10 (1860); Reise der Nov., Lep. ii. p. 289, pl. 36. figs. 9–11 ("1865").   ♂ ♀, Malacca.   This species is evidently very common.

2. ZEMEROS ALBIPUNCTATA (Pl. LXIX. fig. 10), Butler, Cist. Ent. i. p. 236 (1874). ♂ ♀, Malacca.   We have the female of this species from Borneo.   It seems to be a common species.

Genus ABISARA, Felder.

1. ABISARA SAVITRI, Felder, Wien. ent. Mon. iv. p. 397. n. 12 (1860).   ♂ ♀, Malacca.   This is quite distinct from the Indian species *A. susa* of Hewitson.

2. ABISARA KAUSAMBI, Felder, Wien. ent. Mon. iv. p. 397, n. 11 (1860).   ♂ ♀, Malacca.   A common species, the male of which is very different from *A. echerius*.

3. ABISARA HAQUINUS (*Papilio haquinus*), Fabricius, Ent. Syst. iii. 1, p. 55. n. 169 (1793). ♀, Malacca.

Subgenus LAXITA.

4. ABISARA TANITA, (*Taxila tanita*) Hewitson, Ex. Butt. ii. *Tax.* pl. i. (1861). ♀, Malacca. The rounded wings of this and the allied species distinguish them from typical *Abisara*; but I am doubtful of its generic distinction. *T. orphna*, placed by Kirby as type of *Taxila*, must, as I have shown elsewhere, give place to *T. egeon* or *T. fylla*, *T. orphna* having been indicated in the 'Genera of Diurnal Lepidoptera' as *not* the type of the genus.

Family LYCÆNIDÆ.

Subfamily LYCÆNINÆ, Butler.

Genus PORITIA, Moore.

1. PORITIA SUMATRÆ, (*Pseudodipsas sumatræ*) Felder, Reise der Nov., Lep. ii. p. 259, pl. 36. figs. 24–26 (1865). ♀, Malacca; ♂, Penang.

Genus GERYDUS, Boisduval.

1. GERYDUS HORSFIELDI, (*Miletus Horsfieldi*) Moore, Cat. Lep. E.I. C. i. p. 19, pl. 1*a*. fig. 2 (1857). ♂ ♀, Malacca; ♂, Penang.

2. GERYDUS NIVALIS?, (*Miletus nivalis*) Druce, P. Z. S. p. 348. n. 4 (1873). Malacca. I cannot identify this species with certainty; our example measures only 10 lines in expanse of wing.

3. GERYDUS SYMETHUS, (*Papilio symethus*) Cramer, Pap. Exot. ii. pl. 149. figs. B, C (1779). ♂, ♀, Malacca; ♀, Penang.

Genus ANOPS, Boisduval.

1. ANOPS MALAYICA, Felder, Reise der Nov., Lep. ii. p. 221, pl. 28. fig. 18 (1865). ♂ ♀, Malacca.

2. ANOPS SPERTHIS, (*Anops sperthis*) Felder, Reise der Nov., Lep. ii. p. 222 (1865). ♂, Malacca; ♀, Penang.

Genus LIPHYRA, Westwood.

1. LIPHYRA BRASSOLIS, Westwood, Proc. Ent. Soc. p. 31 (1864). ♂, Malacca.

Genus LYCÆNOPSIS, Felder.

1. LYCÆNOPSIS HARALDUS, (*Papilio haraldus*) Fabricius, Maut. Ins. ii. p. 82. n. 744 (1787). ♂, Malacca. This is *L. ananga* of Felder.

Genus CASTALIUS, Hübner.

1. CASTALIUS ROSIMON, (*Papilio rosimon*) Fabricius, Syst. Ent. p. 523. n. 341 (1775). ♂ ♀, Malacca; Penang.

2. CASTALIUS ETHION, (♂ *Lycæna ethion*) Westwood & Hewitson, Gen. Diurn. Lep. pl. 76. fig. 3 (1852). ♀, Above very like *C. roxus*, but with the white band extending nearly to the costa of primaries, internally excavated and externally broadly truncate-clavate at its upper extremity; below white, spotted with dark brown, as in the male. Expanse of wings 11 lines. ♀, Malacca.

3. CASTALIUS ROXUS, (*Polyommatus roxus*) Godart, Enc. Méth. ix. p. 659. n. 142 (1829), Horsfield, Cat. pl. 2. figs. 4, 4 a (1828). ♂ ♀, Malacca.

Genus LAMPIDES, Hübner.

1. LAMPIDES PATALA, (*Lycæna patala*) Kollar, Hügel's Kaschmir, iv. 2, p. 419 (1848). ♂, Malacca.

2. LAMPIDES MACROPHTHALMA, (*Lycæna macrophthalma*) Felder, Verh. zool.-bot. Ges. xii. p. 483. n. 115 (1862). ♂, Malacca.

3. LAMPIDES BEROË, (*Lycæna beroë*) Felder, Reise der Nov., Lep. ii. p. 275, pl. 34. fig. 36 (1865). ♂ ♀, Malacca.

4. LAMPIDES ALUTA, (*Cupido aluta*) Druce, P. Z. S. p. 349. n. 16, pl. xxxii. fig. 8 (1873). ♂ ♀, Malacca. Nearly allied to *L. nora* of Felder, from Amboina, but smaller, with less acuminate primaries.

5. LAMPIDES ALMORA?, (*Cupido almora*) Druce, P. Z. S. p. 349. n. 14, pl. xxxii. fig. 7 (1873). ♂, Malacca.

6. LAMPIDES BÆTICUS, (*Papilio bæticus*) Linnæus, Syst. Nat. i. 2, p. 789 (1767). ♂ ♀, Penang.

7. LAMPIDES CÆRULEA, (*Cupido cærulea*) Druce, P. Z. S. p. 349. n. 13, pl. xxxii. fig. 6 (1873). ♂, Malacca.

8. LAMPIDES AGNATA, (*Cupido agnata*) Druce, P. Z. S. p. 106, pl. xvi. figs. 2–4 (1874). ♂ ♀, Malacca.

9. LAMPIDES PSEUDELPIS, n. sp. (Pl. LXVIII. figs. 7, 8). Nearly allied to *L. elpis*, which it resembles above; below, however, with the white transverse bands narrower, irregular, and broken up, the ground-colour more distinctly bluish opaline, and the large submarginal black spot of secondaries less broadly encircled by ochreous. Expanse of wings 1 inch 6 lines. ♂ ♀, Malacca.

10. LAMPIDES KANKENA?, (*Lycæna kankena*) Felder, Verh. zool.-bot. Ges. xii. p. 481 (1862). ♂ ♀, Malacca.

Genus CATAPÆCILMA, n. gen.

Allied to *Lampides* and *Miletus*, but differs from both in having three tails to the secondaries; the antennæ are long, slender, and acuminate; the wing-cells and neuration generally are very like that of *Lampides*; the wings below are spangled with silver, much as in *Miletus* (*Hypochrysops*, part, Felder).

1. CATAPÆCILMA ELEGANS, (*Hypochrysops elegans*) Druce, P. Z. S. p. 350, pl. xxxii. fig. 12 (1873). ♂, Malacca; Penang. The type was described from Borneo. The figure, which is extremely rough, represents the species with three to four tails to secondaries; but either the fourth has been broken off in all the examples which I have examined, or does not really exist.

Genus LYCÆNA, Fabricius.

1. LYCÆNA CAGAYA, Felder, Reise der Nov., Lep. ii. p. 278, pl. 34. figs. 11–13 (1865). ♀, Malacca.

2. LYCÆNA LYSIZONE, Snellen, Tijd. voor Ent. pl. 7. fig. 2 (1876). ♂♀, Malacca; Penang. Rather larger than *L. karsandra*; paler below.

3. LYCÆNA KARSANDRA, Moore, P. Z. S. p. 505, pl. 31. fig. 7 (1865). ♀, Penang.

4. LYCÆNA SANGRA, (*Polyommatus sangra*) Moore, P. Z. S. p. 772, pl. xli. fig. 8 (1865). ♂, Malacca; ♀, Penang.

Subfamily THECLINÆ, Butler.

Genus AMBLYPODIA, Horsfield.

1. AMBLYPODIA CENTAURUS, (*Papilio centaurus*) Fabricius, Syst. Ent. p. 520. n. 329 (1775). ♂♀, Malacca; ♀, Penang.

2. AMBLYPODIA EUMOLPHUS, (*Papilio eumolphus*) Cramer, Pap. Exot. iv. pl. 299. figs. G, H (1872). ♂♀, Malacca.

3. AMBLYPODIA ANTHELUS, Westwood, Gen. Diurn. Lep. pl. 74. fig. 6 (1852). ♂, Malacca.

4. AMBLYPODIA ADATHA, Hewitson, Cat. Lyc. B. M. pl. 4. figs. 29–31 (1862). ♂♀, Malacca. This is quite distinct from *A. micale.*

5. AMBLYPODIA ATOSIA, Hewitson, Ill. Diurn. Lep. p. 9, pl. 2. figs. 8, 9 (1863). ♂♀, Malacca.

6. AMBLYPODIA AMPHIMUTA, Felder, Wien. ent. Mon. iv. p. 396. n. 6 (1860). ♂♀, Malacca.

7. AMBLYPODIA EPIMUTA, Moore, Cat. Lep. Mus. E.I. C. i. p. 42. n. 62 (1857). ♂♀, Malacca.

8. AMBLYPODIA ANTIMUTA, (*Arhopala antimuta*) Felder, Reise der Nov., Lep. ii. p. 233 (1867). ♀, Malacca.

9. AMBLYPODIA AROA, Hewitson, Ill. Diurn. Lep. p. 13, pl. 2. fig. 12 (1863). ♂, Malacca. A species from India, labelled by Mr. Hewitson as *A. aroa*, is more nearly allied to *A. epimuta.*

10. AMBLYPODIA METAMUTA, ♀, Hewitson, Ill. Diurn. Lep. p. 13, pl. 2. figs. 154, 155 (1863). ♂♀, Malacca. The whole upper surface of the male is coloured deep purple as in the primaries of the female, but without the black border.

11. AMBLYPODIA HYPOMUTA, Hewitson, Cat. Lyc. Brit. Mus. i. p. 11, pl. 6. figs. 63, 64 (1862). ♂ ♀, Malacca; ♀, Penang. Quite distinct from *A. amphimuta.*

12. AMBLYPODIA APIDANUS, (*Papilio apidanus*) Cramer, Pap. Exot. ii. pl. 137. figs. F, G (1779). ♂, Malacca.

13. AMBLYPODIA ? DIARDI, Hewitson, Cat. Lyc. B. M. p. 9, pl. 5. figs. 41, 42 (1862). ♂, Malacca. This species would, I think, be better placed in *Ulica.*

Genus HYPOLYCÆNA, Felder.

1. HYPOLYCÆNA ERYLUS, (*Polyommatus erylus*) Godart, Enc. Méth. ix. p. 633. n. 60 (1823). ♂ ♀, Malacca; ♀, Penang.

2. HYPOLYCÆNA ? ETOLUS, (*Papilio etolus*) Fabricius, Mant. Ins. ii. p. 66 (1787). ♂, Malacca. This species appears to me to agree better with *Myrina.*

Genus APHNÆUS, Hübner.

1. APHNÆUS LOHITA, (*Amblypodia lohita*) Horsfield, Cat. Lep. Mus. E.I. C. p. 106. n. 38 (1829). ♂ ♀, Malacca; Penang.

Genus DEUDORIX, Hewitson.

1. DEUDORIX PETOSIRIS, Hewitson, Ill. Diurn. Lep. p. 22, pl. 9. figs. 30, 31 (1863). ♀, Malacca.

2. DEUDORIX PHERETIMA, Hewitson, Ill. Diurn. Lep. p. 21, pl. 9. figs. 27–29 (1863). ♂, Malacca.

3. DEUDORIX JARBAS, (*Papilio jarbas*) Fabricius, Mant. Ins. ii. p. 68. n. 648 (1787). ♂, Malacca. In *D. jarbas* the veins of the primaries are blackened, which readily distinguishes it from *D. melampus.*

4. DEUDORIX DOMITIA, Hewitson, Ill. Diurn. Lep. p. 19, pl. 6. figs. 6, 7 (1863). ♂, Malacca.

Genus IOLAUS, Hübner.

1. IOLAUS LONGINUS, (*Hesperia longinus*) Fabricius, Ent. Syst. Suppl. p. 430 (1798). ♀, Penang.

Genus MYRINA, Fabricius.

1. MYRINA MEGISTIA (?), Hewitson, Ill. Diurn. Lep. Suppl. p. 5, pl. 3. figs. 77, 78 (1869). Malacca.

2. MYRINA TRAVANA, Hewitson, Ill. Diurn. Lep. p. 38, pl. 17. figs. 59, 60 (1865). ♂ ♀, Malacca. The male of this is like a *Deudorix* on the upper surface, the female like an *Amblypodia.*

3. MYRINA CHITRA, (*Thecla chitra*) Horsfield, Cat. Lep. E.I. C. p. 97, pl. 1. fig. 5 (1829). ♀, Malacca.

4. MYRINA MARCIANA, Hewitson, Ill. Diurn. Lep. p. 34; pl. 12. figs. 12, 13 (1868). ♂, Malacca.

5. MYRINA LAPITHIS, Moore, Cat. Lep. Mus. E.I. C. i. p. 48. n. 79 (1857). ♂, Malacca.

6. MYRINA THARIS, (*Oxylides tharis*) Hübner, Zutr. ex. Schmett. figs. 883, 884 (1837). ♂ ♀, Malacca.

7. MYRINA AMRITA, Felder, Wien. ent. Mon. iv. p. 395 (1860).   ♂ ♀, Malacca.

8. MYRINA FREJA, (*Hesperia freja*) Fabricius, Ent. Syst. iii. 1, p. 263 (1793). ♀, Penang.

Genus LOXURA, Horsfield.

1. LOXURA ATYMNUS, (*Papilio atymnus*) Cramer, Pap. Exot. iv. pl. 331. figs. D, E (1784).   ♂ ♀, Malacca; ♂, Penang.

Family PAPILIONIDÆ.

Subfamily PIERINÆ, Bates.

Genus DELIAS, Hübner.

1. DELIAS METARETE, n. sp.   Nearly allied to *D. hyparete*, but longer in the wing, the greyish apical area of primaries more uniform, in the female with a series of five or six whitish (not yellowish) decreasing streaks between the veins; external border of secondaries pale grey in the male, broadly blackish in the female, the red markings distinctly visible.   Apical spots of primaries below bounded within by a slender angulated greyish line: internal area of secondaries of a much brighter yellow colour; scarlet submarginal spots about twice as large at anal angle, not extending above the second subcostal branch, bounded inwardly by a slender interrupted black streak.   Expanse of wings 3 inches 2 lines.   ♂ ♀, Malacca; Penang.   This species occurs also in Borneo.

2. DELIAS DIONE, (♂, *Papilio dione*) Drury, Ill. Ex. Ent. ii. pl. 8. figs. 3, 4 (1773). ♀, Malacca.   The female resembles that of *D. egialea*, excepting that the bands are white instead of orange.

Genus TERIAS, Swainson.

1. TERIAS FORMOSA, (*Eurema formosa*) Hübner, Zutr. ex. Schmett. figs. 979, 980 (1837).   ♂ ♀, Malacca.

2. TERIAS SARI, Horsfield, Cat. Lep. Mus. E.I. C. p. 136 (1829).   ♂ ♀, Malacca. "Occurs at Province Wellesley" (*W. L. Distant*).

3. TERIAS HECABEOIDES, Ménétriés, Cat. Mus. Petr., Lep. i. p. 85, pl. 2. fig. 2 (1855). ♂ ♀, Malacca; ♂, Penang.

4. TERIAS INANATA, Butler, P. Z. S. p. 617. n. 35 (1875).   ♀, Malacca (3 examples).

5. TERIAS PUMILARIS, Butler, P. Z. S. p. 617. n. 36, pl. lxvii. fig. 7 (1875).   ♂, Malacca. I cannot find any distinguishing characters to separate the two preceding Malayan forms from their representatives in the New Hebrides.

Genus CATOPSILIA, Hübner.

1. CATOPSILIA CROCALE, (*Papilio crocale*) Cramer, Pap. Exot. i. pl. 55. figs. C, D (1779). ♂ ♀, Malacca; ♀, Penang.

2. CATOPSILIA CATILLA, (*Papilio catilla*) Cramer, Pap. Exot. iii. pl. 229. figs, D, E (1782).   ♂ ♀, Malacca ; ♀, Penang.

3. CATOPSILIA CHRYSEIS, (*Papilio chryseis*) Drury, Ill. Ex. Ent. i. pl. 12. figs. 3, 4 (1773).   ♂ ♀, Malacca ; Penang.

4. CATOPSILIA SCYLLA, (*Papilio scylla*) Linnæus, Mus. Lud. Ulr. p. 242 (1764). ♀, Malacca ; ♂, Penang. The specimens are all of the *C. elesia* type, having the secondaries bright golden orange ; one example nearly approaches the brighter specimens of typical *C. scylla*.

Genus HEBOMOIA, Hübner.

1. HEBOMOIA GLAUCIPPE, (*Papilio glaucippe*) Linnæus, Mus. Lud. Ulr. p. 240 (1764). ♂, Malacca.

Genus APPIAS, Hübner.

1. APPIAS PLANA, n. sp.   ♂. Constantly differing from the Javan *A. leptis* of Felder in the absence of the black border of secondaries, a trace only of which exists at the apex of these wings ; also larger, the primaries more produced ; the apex of primaries below and the secondaries of a clearer cream-colour : expanse of wings 2 inches 8 lines. ♂, Malacca. This species is also common in Borneo.

2. APPIAS CARDENA, (*Pieris cardena*) Hewitson, Ex. Butt. ii. *Pier.* pl. 3. figs. 17, 18 (1861).   ♂, Malacca. Also common in Borneo.

3. APPIAS ELEONORA, (*Pieris eleonora*) Boisduval, Sp. Gén. Lép. i. p. 481. n. 64 (1836). ♂ ♀, Malacca ; ♀, Penang. The female nearly resembles that sex of *A. enarete*.

4. APPIAS NATHALIA, (*Pieris nathalia*) Felder, Wien. ent. Monatschr. vi. p. 285 (1862). ♂ ♀, Malacca. Occurs also at Singapore. The type described from the Philippines.

5. APPIAS PANDA, (*Pieris panda*) Godart, Enc. Méth. ix. p. 147, n. 102 (1819). ♂, Malacca. This appears to be only the worn male of *A. nathalia* ; but the black border seems narrower.

6. APPIAS NERO, (*Papilio nero*) Fabricius, Ent. Syst. iii. 1, p. 153. n. 471 (1793). ♂ ♀, Malacca ; ♂, Penang. Occurs also in Java and Siam.

7. APPIAS FIGULINA, (♀ *Pieris figulina*) Butler, Ann. & Mag. Nat. Hist. ser. 3, vol. xx. p. 399, pl. 8. fig. 1 (1867).   ♂ ♀, Malacca. The male is brighter in colour than *A. nero* ♂, and golden yellow, with a greyish streak connecting the first and second branches of primaries, and usually a greyish indistinct nebula across the disk of second-aries. The type of *A. figulina* was from Singapore ; the species occurs also in Borneo.

Genus BELENOIS, Hübner.

1. BELENOIS CYNIS, (*Pieris cynis*) Hewitson, Ex. Butt. iii. *Pier.* pl. 8. fig. 51 (1866). ♂ ♀, Malacca. The form figured by myself in the 'Transactions of the Entomological Society' may, I think, turn out to be distinct.

Subfamily PAPILIONINÆ, Bates.

Genus ORNITHOPTERA, Boisduval.

1. ORNITHOPTERA RUFICOLLIS, n. sp. ♂. Allied to *O. flavicollis*: wings smaller, comparatively narrower; outer margin of primaries more distinctly inarched; collar carmine. Expanse of wings 5 inches 4 lines to 6 inches 1 line. ♂ Malacca.

Unfortunately no female examples have come; the male seems to be not uncommon. Mr. Distant has an example of the female from Province Wellesley. The primaries are of a rather more rounded form and duller black colour than those of the male; the streaks are more grey in colouring; the secondaries have a broad deeply indented marginal border, six large discal spots, the veins and the base black. It was taken on the 21st of February 1869, on the lower part of " Batu Kawan."

Genus PAPILIO, Linnæus.

1. PAPILIO ANTIPHATES, Cramer, Pap. Exot. i. pl. 72. figs. A, B (1779). ♂ ♀, Malacca.

2. PAPILIO SARPEDON, Linnæus, Mus. Lud. Ulr. p. 196 (1764). ♂ ♀, Malacca; Penang.

3. PAPILIO EVEMON, Boisduval, Sp. Gén. Lép. i. p. 234 (1836). ♂ ♀, Malacca.

4. PAPILIO AXION, Felder, Verh. zool.-bot. Ges. xiv. p. 305. n. 224, p. 350. n. 128 (1864). ♂ ♀, Malacca. A variety of this species occurs in which the bands and spots are cream-coloured.

5. PAPILIO BATHYCLES, Zinken-Sommer, Nova Acta Acad. Nat.-Cur. xv. p. 157, pl. 14. figs. 6, 7 (1831). ♂, Malacca. " *P. bathycles* is common round puddles in wet weather " (*W. L. Distant*).

6. PAPILIO ARYCLES, Boisduval, Sp. Gén. Lép. i. p. 231. n. 51 (1836). ♂, Malacca. One example was taken at Province Wellesley by Mr. Distant.

7. PAPILIO AGAMEMNON, Linnæus, Mus. Lud. Ulr. p. 202 (1764). ♂ ♀, Malacca; ♀, Penang.

8. PAPILIO MALAYANUS, Wallace, Trans. Linn. Soc. xxv. p. 59 (1865). ♂ ♀, Malacca; ♂, Penang. A distinct species, like *P. sthenelus* above, but more like *P. erithonius* below. " Excessively common in gardens, August and September " (*W. L. Distant*).

9. PAPILIO DEMOLION, Cramer, Pap. Exot. i. pl. 80. figs. A, B (1779). ♂ ♀, Malacca.

10. PAPILIO DELESSERTII, Guérin, Deless. Souv. Voy. d. l'Inde, ii. p. 68, pl. 17 (1843). ♂, Malacca.

11. PAPILIO CLYTIA, Linnæus, Mus. Lud. Ulr. p. 296 (1764). ♂, Penang.

12. PAPILIO POLYTES, Linnæus, Mus. Lud. Ulr. p. 186 (1764). ♂ ♀, Malacca; ♂, Penang. Var. *P. stichius*, Hübner, Samml. exot. Schmett. (1806-16). ♂ ♀, Malacca. The male of the latter form has the yellow spots well defined on the underside of secondaries, just as the female has a greater amount of red colouring on the under-

side than than typical *P. polytes*. I suspect this variation to be seasonal, and not polymorphic.

13. PAPILIO HELENUS, Linnæus, Mus. Lud. Ulr. p. 185 (1764). ♂ ♀, Malacca.

14. PAPILIO PREXASPES, Felder, Reise der Nov., Lep. i. p. 107, pl. 15. fig. *d* (1865). ♂ ♀ alacca. Not previously in the Museum collection.

15. PAPILIO ISWARA, White, Entom. i. p. 280 (1842). ♂, Malacca; ♂ ♀, Penang.

16. PAPILIO SATURNUS, Guérin, Deless. Souv. Voy. d. l'Inde, ii. p. 70, pl. 18 (1843). ♂, Malacca.

17. PAPILIO ESPERI, n. sp. (Pl. LXVIII. fig. 7.) ♂, *Papilio protenor* (part), Esper [nec Cramer], Ausl. Schmett. pl. 29. fig. 2 (1785–98). ♀. Primaries above grey, with a diffused broad subapical white band, the base, veins, and broad streaks between them black, a red patch at base of discoidal cell; secondaries precisely as in the male, but with a red spot on anal angle; differences below as above : expanse of wings 5 inches 10 lines. ♂ ♀, Malacca; ♀, Penang.

18. PAPILIO MESTOR, (♂, *Papilio androgeos*), Cramer, Pap. Exot. i. pl. 91. figs. A, B (1779). *Iliades mestor*, Hübner, Verz. bek. Schmett. p. 89. n. 931 (1816). ♀. Differs from the female of *P. Esperi* in having the white patch of primaries transferred to the external angle, and the wings longer; below, the differences are the same as in the male. Expanse of wings 6 inches 5 lines. ♂ ♀, Malacca; ♂, Penang.

19. PAPILIO ACHATES, (♀ *Papilio achates*) Cramer, Pap. Exot. ii. pl. 182. figs. A, B (1779). ♂. Wings much more uniform in colour than in the two preceding species, a red basal streak as in *P. androgeos*; the red submarginal spots below much more restricted, as in the female : expanse of wings 5 inches 6 lines. ♂ ♀, Malacca.

Most Lepidopterists have united the three preceding forms in their collections as polymorphic varieties of *P. agenor*, Linnæus, noticing only the differences of the females. I am satisfied that all three (with other forms referred to *P. memnon, agenor*, and *protenor*) are distinct species.

20. PAPILIO VARUNA, White, Entom. i. p. 280 (1842). ♂ ♀ Malacca; Penang. The type was described from Penang, where it is evidently common.

21. PAPILIO DIPHILUS, Esper, Ausl. Schmett. pl. 40 B. fig. 1 (1785–98). ♂ ♀, Malacca; Penang.

Family HESPERIIDÆ.

Genus CASYAPA, Kirby.

1. CASYAPA THRAX, (*Papilio thrax*) Linnæus, Syst. Nat. i. 2, p. 794 (1767). ♂, Malacca.

2. CASYAPA IRAVA (*Hesperia irava*), Moore, Cat. Lep. E.I. C. i. p. 254 (1857). ♂ ♀, Malacca.

Genus HESPERIA, Fabricius.

1. HESPERIA HARISA, (*Ismene harisa*) Moore, Proc. Zool. Soc. p. 782 (1865). ♀, Malacca.

2. HESPERIA VITTA, Butler, Trans. Ent. Soc. p. 498 (1870); Lep. Exot. pl. lix. fig. 9 (1874). ♂, Malacca.

The examples from Malacca have two minute transparent dots placed obliquely beyond the cell of primaries. I have no opportunity of comparing the type at present, and therefore do not know if there is any trace of these dots in the Bornean form.

3. HESPERIA BADRA, (*Goniloba badra*) Moore, P. Z. S. p. 778 (1865). ♂ ♀, Malacca.

Genus COBALUS, Hübner.

1. COBALUS ELIA, (*Hesperia elia*) Hewitson, Trans. Ent. Soc. ser. 3, vol. ii. p. 489. n. 9 (1866). ♂ ♀, Malacca. The type was described from Sumatra.

2. COBALUS CILIATUS, n. sp. Dark chocolate-brown; primaries with two small spots placed obliquely upon the median interspaces, and three dots in an angular line beyond the cell hyaline white; secondaries with a broad abbreviated transverse band from the abdominal margin, and the fringe (excepting at apex) white; body above greyish; head and prothorax shot with golden green; abdomen irregularly white at the sides. Primaries below as above; secondaries crossed by a broad angular white band, its lower half twice as broad as its upper, connected with the white fringe by a white spur upon the second median interspace; body below white, abdomen laterally banded with black lituræ, anus brown. Expanse of wings 1 inch 7 lines. ♀, Malacca.

Genus PAMPHILA, Fabricius.

Section *Gegenes*, Hübner.

1. PAMPHILA ARIA, (*Ismene aria*) Moore, P. Z. S. p. 784 (1865). ♂ ♀, Malacca. The type was from Bengal.

2. PAMPHILA JULIANUS, (*Hesperia julianus*) Latreille, Enc. Méth. ix. p. 763. n. 99 (1823). ♂. Malacca. Originally discovered in Java.

3. PAMPHILA MATTHIAS, (*Hesperia matthias*) Fabricius, Ent. Syst. Suppl. p. 433 (1798). ♂ ♀, Malacca; ♂, Penang.

Section *Pamphila* (typical).

4. PAMPHILA AUGIAS, Linnæus, Syst. Nat. i. 2, p. 794. n. 257 (1767). ♂, Malacca; Penang.

5. PAMPHILA MÆSOIDES, n. sp. Above very like *P. mæsa* of Moore, deep chocolate-brown, banded and spotted with tawny; below similarly marked to *P. mæsa*, but suffused with deep tawny throughout. Expanse of wings 1 inch 1 line. ♂ ♀, Malacca. This species is at once distinguished from *P. mæsa* by the tawny colour of its bands (in *P. mæsa* they are ochreous); the outer band of primaries is also somewhat less irregular; and the ground-colour of the underside is tawny instead of yellow.

6. PAMPHILA MARO, (*Hesperia maro*) Fabricius, Ent. Syst. Suppl. p. 432 (1798). ♂, Malacca; ♂ ♀, Penang.

7. PAMPHILA NIGROLIMBATA, (*Thymelicus nigrolimbatus*) Snellen, Tijd. voor Ent. p. 165, pl. 7. fig. 5 (1876). ♂ 'Malacca.

Genus PLASTINGIA, Butler.

1. PLASTINGIA CALLINEURA, (*Hesperia callineura*) Felder, Reise der Nov., Lep. iii. p. 573, pl. 71. figs. 9, 10 (1867). ♂, Malacca.

Genus CYCLOPIDES, Hübner.

1. CYCLOPIDES SALSALA, (*Nisoniades salsala*) Moore, P. Z. S. 786 (1865). ♂ ♀, Malacca. Previously known from Bengal.

Genus ASTICTOPTERUS, Felder.

1. ASTICTOPTERUS JAMA, Felder, Wien. ent. Mon. iv. p. 401. n. 29 (1860). ♂, Malacca.

2. ASTICTOPTERUS XANITES. (Pl. LXIX. fig. 7.) ♀. *Astictopterus xanites*, Butler, Trans. Ent. Soc. p. 510 (1870). ♂ ♀, Malacca. The type was from Sarawak; the male is rather smaller, and has the orange band of primaries abbreviated and of a deeper colour.

3. ASTICTOPTERUS GEMMIFER, n. sp. Chocolate-brown; primaries with a broad post-median bright orange band; end of cell and apical area of primaries and disk of secondaries spotted, in certain lights, with shining amethyst-coloured spots. Expanse of wings 1 inch 6 lines. ♂ ♀, Malacca (5 examples).

4. ASTICTOPTERUS ARMATUS, Druce, P. Z. S. p. 359. n. 3, pl. xxxiii. fig. 7 (1873). ♀, Malacca. The type was described from Borneo.

5. ASTICTOPTERUS DIOCLES, (*Nisoniades diocles*) Moore, P. Z. S. p. 787 (1865). ♂ ♀, Malacca. The type was from Bengal.

6. ASTICTOPTERUS SINDU, Felder, Wien. ent. Mon. iv. p. 401. n. 30 (1860). ♂, Malacca. In this species the orange band is transverse, which at once distinguishes it from the male of *A. xanites*. It is also smaller.

7. ASTICTOPTERUS STELLIFER, n. sp. Above much like a small *A. jama*, deep chocolate-brown. Below irrorated, particularly at apex of primaries, with bronzy golden scales; internal area of primaries whity-brown; a small white spot at the end of the cell in all the wings, and a small dull white spot close to the middle of the submedian vein of secondaries; body below pale brown; palpi clothed with golden hair-scales; antennæ black above, yellow below. Expanse of wings 1 inch. ♂, Malacca (2 examples). The under surface of this species reminds one of *Cyclopides salsala* of Moore.

Genus PLESIONEURA, Felder.

1. PLESIONEURA FOLUS, (*Papilio folus*) Cramer, Pap. Exot. i. pl. 74. fig. F (1779). ♂, Malacca ; Penang.

2. PLESIONEURA ALYSOS, Moore, P. Z. S. p. 789 (1865).   ♂ ♀, Malacca.

3. PLESIONEURA ASMARA (Moore in litt.).   Similar to *P. dan*, but not tawny-tinted, the three spots in the centre of primaries united, and hyaline white : expanse of wings 1 inch 8 lines.   ♂ ♀, Malacca (5 examples).

4. PLESIONEURA DAN, (*Papilio dan*) Fabricius, Mant. Ins. ii. p. 88. n. 798 (1787). ♂, Malacca.

5. PLESIONEURA PINWILLI, n. sp.   (Pl. LXVIII. fig. 4.)   Primaries black with a bluish shot ; a broad oblique shining pale-yellow subhyaline patch, separated by the median nervure and its second and third branches into three spots ; two small spots of the same colour, placed obliquely below it, on the interno-median interspace : secondaries bright orange, the base and the apical portion of external border chocolate-brown ; remainder of outer border, a rounded spot at end of cell, a second near anal angle, and five, sub-marginal, touching the outer border, black : head and thorax greenish grey, vertex of head edged with sordid white ; abdomen orange banded with black.   Primaries below as above, excepting that there is a bifid whitish spot above the end of the cell, a whitish spot at base of interno-median area, and that the inner margin is brown : secondaries bright orange ; the costal and outer borders irregularly purplish black ; fringe brown-ish ; a subcostal dash, a rounded spot at the end of the cell, and a reniform spot near the anal angle black : body below and legs bright ochreous, palpi pale ochreous ; neck below white : antennæ black above, testaceous below.   Expanse of wings 2 inches 2 lines. ♂, Malacca.   Most nearly allied to *P. labrica* of Hewitson from Darjeeling.

Genus TAGIADES, Hübner.

1. TAGIADES RAVI, (*Pterygospidea ravi*) Moore, Proc. Zool. Soc. p. 779 (1865). ♂ ♀, Malacca.   I have compared this with the type, with which it entirely agrees. I have seen specimens of what is evidently a nearly allied species labelled as T. *ravi* ; they differ in having the hind borders of the secondaries above, and the abdominal half below, white.

2. TAGIADES GANA, (*Pterygospidea gana*) Moore, P. Z. S. p. 780 (1865).   ♂ ♀, Malacca. Originally discovered at Bengal.

3. TAGIADES CALLIGANA, n. sp.   (Pl. LXIX. fig. 11.)   Closely allied to *T. atticus*, but the secondaries much narrower, and the two black spots towards anal angle wanting, leaving only two submarginal black spots ; submarginal spots below contiguous, almost forming a border.   Expanse of wings 1 inch 8 lines.   ♂ ♀, Malacca (2 examples).

4. TAGIADES LAVATA, n. sp. (Pl. LXIX. fig. 8.) Wings above dark smoky brown; primaries with a sinuous series of five subapical hyaline white dots; secondaries with anal half of external border snow-white, its inner edge sharply defined; body brown. Primaries below paler; secondaries snow-white, with the costal area brown, its inner edge diffused; a small brown litura across the first median branch; body below whitish. Expanse of wings 1 inch 6 lines. ♂, Malacca. Allied to *T. pralaya*.

HETEROCERA.

Family SPHINGIDÆ.

Subfamily MACROGLOSSINÆ (Grote), Butler.

Genus MACROGLOSSA, Ochsenheimer.

1. MACROGLOSSA PROXIMA, Butler, P. Z. S. p. 4, pl. 1. fig. 1 (1875). ♂, Malacca. The figure of this species in the 'Proceedings' is not satisfactory, the yellow band being left too broad, the diverging basal black streaks being also omitted.

Family ZYGÆNIDÆ.

Subfamily ZYGÆNINÆ, Butler.

Genus SYNTOMIS, Ochsenheimer.

1. SYNTOMIS BASIFERA, Walker, Journ. Linn. Soc. vi. p. 92. n. 37 (1862). ♀, Penang. This species is intermediate in character between *S. apicalis* and *S. detracta*.

Family ARCTIIDÆ.

Genus SPILOSOMA, Stephens.

1. SPILOSOMA, n. sp. A wholly white species. It is in very bad condition; and therefore I prefer not to name it. ♀, Penang.

Family LITHOSIIDÆ.

Subfamily LITHOSIINÆ.

Genus ARGINA, Hübner.

1. ARGINA CRIBRARIA, (*Phalæna cribraria*) Clerck, Icones, pl. 54. fig. 4. ♀, Penang.

Subfamily HYPSINÆ, Butler.

Genus HYPSA, Hübner.

1. HYPSA HELICONIA, (*Phalæna-Noctua heliconia*) Linnæus, Syst. Nat. i. 2, p. 839 707). ♀, Malacca.

2. HYPSA SUBSIMILIS, Walker, Lep. Het. Suppl. i. p. 212 (1864). ♀, Malacca; ♂, Penang.

Subgenus DAMALIS, Hübner.

3. HYPSA (DAMALIS) EGENS, Walker, Lep. Het. ii. p. 453. n. 12 (1854). ♂, Malacca.

Genus NEOCHERA, Hübner.

1. NEOCHERA MARMOREA, (*Hypsa marmorea*) Walker, Lep. Het. vii. p. 1674 (1856). ♀, Penang.

Genus CALLIDULA, Hübner.

1. CALLIDULA ABISARA, n. sp. Nearly allied to *C. sakuni* from Java, but smaller, duller in colour; the band of primaries narrower, more regular, and rounded at its lower extremity; wings below more densely reticulated with brown, and duller. Expanse of wings 1 inch 2 lines. ♂, Malacca (1 example).

Apparently the same as the Bornean form; but the type is not in good condition. Our Bornean examples differ from *C. sakuni* in the colour of the band of primaries, which is yellow with only the inner edge orange. This is probably the case with fresh examples from Malacca.

Genus CLEOSIRIS, Horsfield.

1. CLEOSIRIS CATAMITA, (*Tetragonus catamitus*) Hübner, Zuträge, figs. 653, 654 (1832). ♂, Penang.

Family NYCTEMERIDÆ.

Genus NYCTEMERA, Hübner *.

1. NYCTEMERA TRIPUNCTARIA, (*Phalæna tripunctaria*) Linnæus, Syst. Nat. i. 2, p. 864 n. 226 (1767). ♂, Malacca.

2. NYCTEMERA COLETA (*Phalæna coleta*), Cramer, Pap. Exot. iv. p. 153, pl. 368. fig. II (1782). ♂, Malacca.

Genus SECUSIO, Walker.

1. SECUSIO MUNDIPICTA, (*Nyctemera mundipicta*) Walker, Journ. Linn. Soc. iv. p. 184, n. 7 (1860). ♂, ♀, Malacca; ♀, Penang. The type was from Singapore. I have referred the above species, with *N. trita* (its nearest ally), *N. plagiata*, *N. annulata*, and *N. distincta*, to the genus *Secusio*; they differ from the species of *Nyctemera* in the Museum.

Family CHALCOSIIDÆ.

Genus MILIONIA, Walker.

1. MILIONIA BASALIS, Walker, Lep. Het. Brit. Mus. ii. p. 365 (1854). ♀, Malacca The bands, in the single example sent, are rather wider than in females from Java.

---

* Type *N. coleta*, the first three species belong to *Otroeda*, the fifth to *Amnemopsyche*, the seventh to *Secusio*; the two remaining forms are congeneric..

Genus EUSCHEMA, Hübner.

1. EUSCHEMA SUBREPLETA, Walker, Lep. Het. ii. p. 406. n. 3 (1854). *Hazis bello-naria*, Guénée, Uran. et Phal. ii. pl. 18. fig. 1 (1857). ♂ ♀, Malacca.

Genus AMESIA, Westwood.

1. AMESIA JUVENIS, n. sp. Most nearly allied to *A. venusta*: primaries rich purplish chocolate; apical area, and a series of five marginal spots ultramarine blue; a series of eight submarginal white dots; a sickle-shaped series of five discal white dots: secondaries red-brown, with two submarginal series of white spots, the inner or discal series ill defined: body blue-black; head and tegulæ spotted with blue. Wings below red-brown, costal half of primaries purplish; base of costa metallic green; two submarginal series of white spots uniting into streaks at anal angle of secondaries, the inner series of primaries deeply incurved, the three uppermost spots of the outer series edged internally with blue; body below metallic green spotted with white. Expanse of wings 2 inches 9 lines. ♂, Malacca. Easily distinguished from its allies by its pale red-brown secondaries.

2. AMESIA PEXIFASCIA, Butler, Journ. Linn. Soc. xiii. p. 115. Primaries rich purplish chocolate, external two fifths covered by a broad externally deeply digitate dentated snow-white band, interrupted by the black nervures, the portion crossing the end of the cell divided through the centre into two large spots; secondaries almost exactly as in *A. euplœoides* of Herrich-Schäffer, but the anal angle dark greenish-grey; body deep purplish chocolate, the three last abdominal segments blue-green; palpi, tegulæ, and thorax spotted with lilacine dots. Primaries below as above, but with a basal white spot, two costal, two subcostal, two discoidal, and two interno-median blue-edged white spots; secondaries below nearly as in *A. euplœoides*, but the anal angle grey; body below brown, with a lateral series of blue-edged white spots. Expanse of wings 3 inches 11 lines. ♀, Malacca. The most striking species in the genus.

Genus LAURION, Walker.

1. LAURION CORCULUM, n. sp. Sepia-brown; primaries crossed by an oblique narrow sulphur-yellow band, beyond which the apical area is almost black. Wings below paler, greyish towards the base; primaries with the costa as far as the band blue; secondaries with four apical marginal steel-blue spots; sides of pectus and anterior coxæ blue; venter banded with white. Expanse of wings 1 inch 1 line. ♂, Malacca.

Genus CHALCOSIA, Hübner.

1. CHALCOSIA COLIADOIDES (var. LATIFASCIATA). *Chalcosia coliadoides*, Walker, Journ. Linn. Soc. vi. p. 97 (1862). ♀, Malacca. Differs from the Bornean form in the more regular, non-lunated, broader, subapical, curved, cream-coloured band of primaries.

Genus CYCLOSIA, Hübner.

1. CYCLOSIA PANTHONA, (*Phalæna [Geometra] panthona*) Cramer, Pap. Exot. iv. p. 68, pl. 322. fig. C (1782). ♀, Malacca.

Family LIPARIDÆ.

BIRNARA, n. gen.

Allied to *Pantana*: wings rounded, semitransparent; costal nervure of primaries terminating above the end of the cell; subcostal with four branches, the third and fourth forming a fork to apex; radials emitted near together from the upper part of the cell; discocellulars zigzag; median nervure with four branches, the third and fourth emitted near together: secondaries pyriform, costal nervure extending to apex; subcostal emitting two branches from a short footstalk; radial forming a fourth branch of the median; discocellulars in an oblique slightly angulated line. Type *B. nubila*, n. sp.

1. BIRNARA NUBILA, n. sp. Primaries smoky grey, the veins white, base ochreous; secondaries greyish or sordid white; body ochreous, antennæ brown. Wings below pale grey, becoming smoky grey towards the outer margin; body ochreous. Expanse of wings 2 inches 1 line. ♀, Malacca. Allied to *Pantana bicolor* from China, which, with *Eloria marginalis* and *Genusa discifera*, may be referred to this genus.

Genus KETTELIA, n. gen.

Wings semihyaline, broad: primaries subtriangular; costal nervure extending to beyond the end of the cell; subcostal emitting five branches, the first branch emitted before the end of the cell, the second some distance beyond the end, immediately beyond which the vein forks, the upper ray proceeding to near apex, and then forking again, the lower ray simple; upper radial emitted from the upper part of the discocellular line (which is zigzag); lower radial forming a fourth median branch: secondaries pyriform; costal nervure proceeding to apex; subcostal emitting two branches from the superior extremity of the cell; radial emitted from the inferior extremity of the cell on a level with the third median; upper discocellular acutely angular, its lower portion twice as long as its upper, the lower extremity of the cell is consequently greatly in advance of the upper: body slender; antennæ rather short, broadly pectinated. Type *K. Lowii*.

1. KETTELIA LOWII, n. sp. Semitransparent white, the borders and veins opaque; a pale brown straight line from the costa to the external angle of primaries, and from the apex to the anal angle of secondaries; body white; palpi head, back of tegulæ, base of primaries, sides of abdomen (more or less), and the greater portion of the legs ochreous yellow; antennæ pale brown. Expanse of wings 2 inches 5 lines. ♂, Malacca; Sarawak (*Low*).

Our Malacca example being rubbed, I have taken as my type one of the three specimens kindly presented to the Collection by Mr. Low. The genus is allied to *Penora*.

Genus ORGYIA, Ochsenheimer.

1. ORGYIA TURBATA, n. sp. Primaries ochreous, densely irrorated with ferruginous; the apex, a spot near external angle, and an ill-defined subbasal spot bright golden yellow; subapical area red-brown; two transverse sinuated blackish lines, widely

diverging to costa; a black-edged reniform ferruginous spot at end of cell; a subbasal transverse dusky line: secondaries reddish brown. Wings below sandy brownish; primaries with a large dusky patch over the end of the cell. Expanse of wings 11½ lines. ♂, Malacca. I have seen a second example from Moulmein in Mr. Atkinson's collection.

Genus PARASA, Moore.

1. PARASA BANDURA, Moore, Cat. Lep. E. I. C. ii. p. 417. pl. xi.a fig. 9 (1858–9). ♂, Malacca.

Family SATURNIIDÆ.

Genus ATTACUS, Hübner.

1. ATTACUS ATLAS, (*Phalæna-Bombyx Attacus atlas*) Linnæus, Syst. Nat. p. 808. n. 1. ♂ ♀, Malacca.

Family HYBLÆIDÆ.

Genus HYBLÆA, Fabricius.

1. HYBLÆA CONSTELLATA, Guénée, Noct. ii. p. 391. n. 1261. ♂ ♀, Malacca.

Family OPHIDERIDÆ.

Genus OPHIDERES, Boisduval.

1. OPHIDERES FULLONIA, (♂, ——*fullonia*) Clerck, Icones, tab. 42. figs. 3, 4 (1759–64); *Ophideres cajeta*, Walker (nec Cramer), Lep. Het. xiii. p. 1223 (1857); ♀, *Phalæna-Noctua pomona*, Cramer, Pap. Exot. ii. p. 122, pl. 77. fig. C (1779). ♂ ♀, Malacca.

Clerck gives three figures on pl. 42, and names the central one *O. fullonia*; it may or may not be the *O. fullonia* of Linnæus. I have little doubt that the true *O. cajeta* is identical with Walker's *O. bilineosa*. According to the example in the Banksian cabinet (which, however, is not the type), the *O. dioscoreæ* of Fabricius is a variety of this species in which the silvery spot of primaries is developed into a streak between the median branches.

Family OMMATOPHORIDÆ.

Genus NYCTIPAO, Hübner.

1. NYCTIPAO CREPUSCULARIS, (*Phalæna-Attacus crepuscularis*) Linnæus, Mus. Lud. Ulr. p. 378; Clerck, Icones, pl. 53. figs. 3, 4 (nec 1, 2). Malacca.

Figures 1 and 2 of Clerck's plate represent Guénée's *N. leucotænia*. "Comes into the house at night; I have never known it to be taken in any other way " (*W. L. Distant*).

Family BENDIDÆ.

Genus HULODES, Guénée.

1. HULODES CARANEA, (*Phalæna-Noctua caranea*) Cramer, Pap. Exot. iii. p. 140, pl. 269. figs. E, F (1782). Malacca.

Family THERMESIIDÆ.

Genus COTUZA, Walker.

1. COTUZA DREPANOIDES, Walker, Lep. Het. xv. p. 1552 (1858).   ♂, Penang.

Family URANIIDÆ.

Genus NYCTALEMON, Dalman.

1. NYCTALEMON DOCILE, n. sp.   Allied to *N. hector*, from which it differs in its superior size, deeper and redder coloration, narrower central white band, which is bordered internally by a broad brown band, and externally by a narrow one; the costal border not spotted with white; a broad oblique diffused discal brown band: secondaries with the apical area much less streaked with black, the anal area much more streaked; the marginal interrupted black border narrower; the tails considerably longer and narrower towards their extremities.   Expanse of wings ♂ 5 inches 6 lines, ♀ 6 inches 4 lines. ♂ ♀, Malacca (6 examples).   This species is a well-marked local representative of *N. hector*.

Family URAPTERIDÆ.

Genus URAPTERYX, Leach.

1. URAPTERYX MARGINIPENNIS, n. sp.   Wings greyish brown, transversely streaked with clay-colour, and reticulated throughout with transverse dark-brown hatchings: primaries with the apex broadly, and the outer margin rather narrowly, nearly sulphur-yellow; costa whitish; a yellow spot on the second median interspace: secondaries with the outer margin rather narrowly sulphur-yellow; two decreasing submarginal black dots above the angle of the wing: body greyish brown.   Wings below grey, spot and borders paler than above, reticulations obsolete; body below whitish.   Expanse of wings 2 inches.   ♂, Malacca.

This curious species is not nearly allied to any species with which I am acquainted; it approaches most nearly to *U. crocopterata*; but the wings are very slightly angulated.

Family GEOMETRIDÆ, Guénée.

Genus AGATHIA, Guénée.

1. AGATHIA DISCRIMINATA, Walker, Lep. Het. xxii. p. 591 (1861).   ♂, Penang.

Family ACIDALIIDÆ.

Genus ACIDALIA.

1. ACIDALIA EXCLUSA, (*Hemerophila? exclusa*) Walker, Cat. Lep. Het. xxi. p. 320 (1860) ; *Macaria obstataria*, Walker, *l. c.* xxiii. p. 928 (1861).   ♂, Malacca.   Walker notes this species as Bremer's *Philobia cinerearia*; but, as I have no Chinese examples before me, I cannot certify the correctness of this determination.

Genus ZANCLOPTERYX, Herrich-Schäffer.

1. ZANCLOPTERYX SAPONARIA, Guénée, Phal. ii. p. 16. n. 914 (1857).   Malacca.

Family MICRONIIDÆ.

Genus MICRONIA, Guénée.

1. MICRONIA ASTHENIATA, Guénée, Phal. ii. p. 24. n. 925 (1857).   Malacca.

Family ZERENIIDÆ.

Genus GENUSA, Walker.

1. GENUSA BIGUTTA, Walker, Cat. Lep. Het. iv. p. 818. n. 1 (1855).   ♂, Malacca.
G. delineata and *circumdata* must be retained among the *Liparidæ*; they are not in any way related to *G. bigutta* (which I consider the type of the genus).

Family MARGARODIDÆ.

Genus GLYPHODES, Guénée.

1. GLYPHODES AMETHYSTA, n. sp.   Basal two thirds of wings hyaline white; external third purplish pearly, bounded internally by a sinuated black line; primaries with a tapering bronze band from costa to inner margin at basal third, a subquadrate spot of the same colour at the end of the cell, costa brown-bordered, a bronze brown discal diffused tapering streak; body probably brown, but too much rubbed for description. Expanse of wings 1 inch 2 lines.   ♂, Malacca.

Family BOTIDIDÆ.

Genus BOTYS, Latreille.

1. BOTYS PTEROPHORALIS, Walker, Cat. Lep. Het. Suppl. iv. p. 1413 (1865).   ♂, Malacca.
2. BOTYS, sp. — ?   Closely allied to *B. contigualis*.   In bad condition.   Malacca.

Therefore it appears that of the 280 species of determinable Lepidoptera collected by Captain Pinwill, 41 are new or hitherto unnamed forms*.

---

* The following Table includes the whole of the species hitherto recorded from Malacca; the species noted in this paper from Penang only are necessarily omitted.   The names of some of the genera formerly included under *Euplœa* have been altered subsequent to the reading of this paper.   Consult Journ. Linn. Soc. Zool. vol. xiv. p. 290 (1878), "On the Butterflies in the Collection of the British Museum hitherto referred to the Genus *Euplœa* of Fabricius."

Tabular View of the Butterflies of Malacca.

Species from Malacca.	Assam & Nepal.	Moulmein.	Ceylon.	Penang.	Singapore.	Borneo.	Sumatra.	Java.	Siam.	China.	South Pacific.	Australia.	Remarks.
Euplœa phœbus, *Butler*	..	..	..	*	*	*	..	*	*				
„ Oehsenheimeri, *Lucas*	*	..	..	*	*	*	..	*	*				
„ Grotei, *Felder*	..	..	..	..	..	..	..	..	..	..	..	..	Type from Cochin.
„ margarita, *Butler*	..	..	..	*	..	*	..	*	*				
„ chloë, *Guérin*	..	..	..	..	..	..	*						
„ Bremeri, *Felder*	*	..	..	*	*	*	*						
„ Ménétriési, *Felder*	*	..	..	..	..	..	..	*		*			
„ Pinwilli, *Butler.*													
„ midamus, *Linn.*	*	..	..	*	*	*	..	*					
Calliplœa diocletianus, *Fabricius*	*	..	..	*	*	*	..	*	*				
„ vestigiata, *Butler*	..	..	..	..	..	..	..	*	*				
„ Lederori, *Felder*	..	..	..	..	..	..							
Salpinx leucogonis, *Butler.*	*												
Danais plexippus, *Linn.*	*	..	*	*	..	..	..	*	*	*			
„ melanippus, *Cramer*	*	..	..	*	..	..	*	*	*	*			
„ vulgaris, *Butler*	*	*	..	..	*	*	..	*					
„ septentrionis, *Butler*	*	..	*	*	*								
„ grammica, *Boisduval*	..	*	..	..	..	..	..	*					
„ melaneus, *Cramer*	*	..	..	*									
„ crocea, *Butler*	*	..	..	*	*	*	..	*					
Ideopsis daos, *Boisduval*	..	..	..	*	*	*	*	*	..	..			
Hestia lynceus, *Drury*	..	..	..	*	..	..	..	*	*				
„ linteata, *Butler.*													
Melanitis leda, *Linn.*	*	*	*	*	..	..	*	*	*	*			
Lethe europa, *Fabricius*	*	..	..	*	..	*	..	*	*				
Cœlites humilis, *Butler.*													
Mycalesis mineus, *Hewitson*	..	..	..	*	*	*	*						
„ polydecta, *Cramer*	..	..	..	*	*	*	..	*					
„ janardana, *Moore*	..	..	..	..	*	..	..	*					
„ nautilus, *Butler*	*	..	..	..	..	..	..						
„ hesione, *Cramer*	..	..	..	*	..	*	..	*					
„ fusca, *Felder*	..	..	..	..	..	*							
Ypthima philomela, *Hübner*	..	..	..	*									
„ methora, *Hewitson*	..	..	..	*									
„ corticaria, *Butler.*													
Elymnias nigrescens, *Butler*	..	..	..	..	*	*							
„ lutescens, *Butler*	..	..	..	*	*	*	..						
„ melida, *Hewitson*	..	..	..	*	*								
Amathusia phidippus, *Linn.*	..	..	..	*	..	*	..	*	*				
Zeuxidia amethystus, *Butler*	..	..	..	..	*	*							
Discophora menetho, *Fabricius*	*	..	..	..	..	*	..	..	..	*			
„ tullia, *Cramer*	*	..	..	..	..	*	..	..	..		*		
Xanthotænia busiris, *Westwood*	..	..	..	..	*	*	*						
Clerome gracilis, *Butler*	..	..	..	..	*	*	*						
„ faunula, *Westwood*	..	..	..	..	*	*							
Thaumantis noureddin, *Westwood*	..	..	..	*	..	*							
„ pseudaliris, *Butler.*													
Doleschallia pratipa, *Felder*	..	..	..	*									
Charaxes jalysus, *Felder*	..	..	..	*	*	..	*	*					
„ hebe, *Butler*	..	..	..	..	..	*	*						
„ Schreiberi, *Godart*	..	..	..	..	..	..	*	..	*				
Prothoe caledonia, *Hewitson*	..	..	..	..	..	..	*						
Adolias adonia, *Cramer*	..	..	..	..	..	..	..	..	*				
„ garuda, *Moore*	..	..	..	..	..	..	..	*	*				
„ jama, *Felder*	..	*	..	..	..	..							

Species from Malacca.	Assam & Nepal.	Moulmein.	Ceylon.	Penang.	Singapore.	Borneo.	Sumatra.	Java.	Siam.	China.	South Pacific.	Australia.	Remarks.
Adolias laverna, *Butler*	..	..	.	*									
„ ramada, *Moore*	..	..	.	..	..	*							
„ asoka, *Felder.*													
„ monina, *Fabricius*	..	..	..	*	*								
„ ludekingii, *Vollenhoven*	..	..	..	..	..	..	*						
Symphœdra dirtea, *Fabricius*	..	..	..	*	..	*	*	*					
„ ? cunalea, *Guérin.*													
Tanaëcia aruna, *Felder*	..	..	..	..	..	..	*						
„ pulasara, *Moore*	..	..	..	*	*	..	*						
Athyma larymna, *Westwood*	*	*	..	*	..	*	..	*					
„ leucothoë, *Linnæus*	*	..	..	*	..	..	..	*	*	*			
„ kresna, *Moore*	..	..	..	*	*	*	*						
„ sabrata, *Moore*	..	..	..	..	..	*							
„ nivifera, *Butler*	*	..	..	..	..	*							
„ ambara, *Druce*	..	..	..	..	..	*							
„ clerica, *Butler.*													
„ idita, *Moore*	*	..	..	..	*	..	*						
„ pravara, *Moore*	..	..	..	*	*	*	*	*					
„ urvasi, *Felder.*													
Euripus euplœoides, *Felder.*													
Eulacura osteria, *Westwood*	*	..	..	*	*								
Lebadea alankara, *Horsfield*	..	..	..	..	*	*							
„ martha, *Fabricius*	..	..	..	..	*	*	..	*					
Limenitis procris, *Cramer*	*	..	..	*	*	*							
Neptis mamaja, *Butler*	..	..	..	*									
„ leuconata, *Butler.*													
„ gononata, *Butler.*													
„ nata, *Moore*	*	..	..	..	*	*							
„ vikasi, *Horsfield*	*	..	..	..	*	*	*						
„ columella, *Cramer*	*	..	..	..	..	..	..	..	*				
„ dorelia, *Butler*	..	..	..	..	*								
„ tiga, *Moore*	..	..	..	*	*	*	*						
„ paraka, *Butler.*													
„ dindinga, *Butler*	..	*											
„ hordonia, *Stoll*	*	..	*	..	*	*	..	*	*	*	*	*	Probably more than one species.
Pandita sinope, *Moore*	*	..	..	*	*								
Hypolimnas incommoda, *Butler.*													
Cethosia hypsina, *Felder*	*												
Cirrochroa orissa, *Felder*	*	..	..	*	..	*							
„ johannes, *Butler.*													
„ bajadeta, *Moore*	..	..	..	..	..	..	*						
„ rotundata, *Butler.*													
Terinos teuthras, *Hewitson*	..	..	..	*	*								
„ robertsia, *Butler*	..	..	..	*									
Messaras erymanthis, *Drury*	*	*	..	*	..	*	..	*	*	*			
Atella phalanta, *Drury*	*	*	*	*	..	*	..	*	*				
„ sinha, *Kollar*	*	*	..	*	..	*	..	*	*				
Cynthia erotella, *Butler*	..	..	..	*	..	*	..	*					
Parthenos lilacinus, *Butler*	*	..	..	*	..	*							
Cyrestis rabria, *Moore*	..	*	..	*	..	*	..	*					
„ nivea, *Zinken-Sommer*	*	*	..	*	..	*	..	*	..	..	..	..	Probably Felder's *C. nivalis.*
Ergolis ariadne, *Linnæus*	*	..	*	*	..	*	..	*	*				
„ merione, *Cramer*	*	..	..	*	..	..	..	*					
Junonia iphita, *Cramer*	*	..	*	*	..	*	..	*					
„ laomedia, *Linnæus*	*	*	*	*	..	*	..	*	*				
„ lomonias, *Linnæus*	*	*	*	*	..	..	..	*	*	*			
„ orithya, *Linnæus*	*	*	*	*	..	*	..	*	..	*			
„ asterie, *Linnæus*	*	..	*	*	..	..	..	*	*	*			
„ eudoxia, *Guérin.*													

Species from Malacca.	Assam & Nepal.	Moulmein.	Ceylon.	Penang.	Singapore.	Borneo.	Sumatra.	Java.	Siam.	China.	South Pacific.	Australia.	Remarks.
Rhinopalpa fulva, *Felder.*													
Zemeros emesoides, *Felder*						*							
,, albipunctata, *Butler*						*							
Albisara savitri, *Felder*					*	*							
,, kausambi, *Felder*					*	*	*	*					
,, haquinus, *Fabricius*	*				*			*					
,, tanita, *Hewitson*					*	*							
,, dumajanti, *Felder.*													
Poritia sumatræ, *Felder*													
Gerydus Horsfieldi, *Moore*				*	*	*	*	*					Possibly more than one species here.
,, nivalis, *Druce*				*	*	*							
,, symethus, *Cramer*				*	*								
Anops malayica, *Felder.*													
,, sperthis, *Felder*				*									
Liphyra brassolis, *Westwood.*													
Lycænopsis haraldus, *Fabricius*							*						
Castalius rosimon, *Fabricius*	*	*	*	*		*		*	*				
,, ethion, *Westwood*		*	*	*				*	*				
,, elna, *Hewitson*								*					
,, roxus, *Godart*						*		*					
Lampides patala, *Kollar*			*										Type from N. India.
,, macrophthalma, *Felder*		*	*									*	Type from the Philippines.
,, beroë, *Felder*				*									
,, aluta, *Druce*						*							
,, almora, *Druce*						*							
,, cærulea, *Druce*				*									
,, agnata, *Druce*						*			*				
,, pseudelpis, *Butler.*													
,, kankena, *Felder*			*										Type from the Nicobars. Also from Southern India.
Catapœcilma elegans, *Druce*			*	*		*							Type from the Philippines.
Lycæna cagaya, *Felder*				*					*				Type from Bengal.
,, lysizone, *Snellen*				*									
,, sangra, *Moore*		*	*	*					*				Type from Bengal.
Amblypodia centaurus, *Fabricius*		*	*	*			*		*				
,, eumolphus, *Cramer*		*		*		*		*					
,, anthelus, *Westwood*		*											
,, adatha, *Hewitson*					*	*							
,, atosia, *Hewitson*					*		*						
,, amphimuta, *Felder*						*							
,, epimuta, *Moore*	*												Type from India.
,, antimuta, *Felder.*													
,, aroa, *Hewitson*							*						
,, metamuta, *Hewitson*							*						
,, hypomuta, *Hewitson*						*							
,, apidanus, *Cramer*		*						*					
,, Diardi, *Hewitson*									*				
,, vihara, *Felder.*													
,, inornata, *Felder.*													
Hypolycæna erylus, *Godart*	*	*		*									
,, etolus, *Fabricius*	*	*		*									
Amphnæus lohita, *Horsf.*		*		*					*	*			
Deudorix petosiris, *Hewitson*		*							*				Also from Silhet.
,, pheretima, *Hewitson*					*	*							
,, jarbus, *Fabricius*					*			*					
,, domitia, *Hewitson*						*							
Myrina megistia, *Hewitson.*													
,, travana, *Hewitson*						*	*						
,, chitra, *Horsf.*					*	*		*					
,, marciana, *Hewitson*					*		*						

Species from Malacca.	Assam & Nepal.	Moulmein.	Ceylon.	Penang.	Singapore.	Borneo.	Sumatra.	Java.	Siam.	China.	South Pacific.	Australia.	Remarks.
Myrina lapithis, *Moore*		*			*	*		*					
" tharis, *Hübner*	*					*		*					
" amrita, *Felder*	*												
" anasuja, *Felder*.													
Loxura atymnus, *Cramer*		*	*	*			*	*	*	*			Type from Coromandel.
Dolias metarote, *Butler*	*			*	*	*		*					
" dione, *Drury*					*				*				
" ninus, *Wallace*.													
Torias formosa, *Hübner*	*					*							
" sari, *Horsfield*				*				*					
" hecabeoides, *Ménétriés*	*		*	*									
" inanata, *Butler*												*	
" pumilaris, *Butler*												*	
" senna, *Felder*.													
Catopsilia crocale, *Cramer*	*	*	*	*		*		*	*	*		*	
" catilla, *Cramer*	*	*	*	*			*	*		*		*	
" chryseis, *Drury*	*	*						*		*		*	
" scylla, *Linnæus*								*				*	
Hebomoia glaucippe, *Linnæus*	*	*	*					*		*			Possibly more than one species.
Appias plana, *Butler*					*								
" cardena, *Hewitson*					*								
" cleonora, *Boisd.*		*		*	*								
" nathalia, *Felder*					*								Type from the Philippines.
" panda, *Godart*								*					
" nero, *Fabricius*				*	*		*	*	*				
" figulina, *Butler*				*	*								
Belenois cynis, *Hewitson*						*							
Ornithoptera ruficollis, *Butler*													
Papilio antiphates, *Cramer*	*			*		*						*	
" sarpedon, *Linn.*	*	*		*	*								
" evemon, *Boisd.*	*	*		*	*								
" axion, *Felder*	*	*		*	*		*						
" bathycles, *Zinken-Sommer*	*			*	*	*		*					
" arycles, *Boisd.*	*			*	*	*	*						
" rama, *Felder*.													
" agamemnon, *Linn.*	*			*			*	*					
" malayanus, *Wallace*				*				*					
" demolion, *Cramer*		*			*	*	*						An allied form came from S. India.
" Doleschallii, *Guérin*						*	*	*					
" polytes, *Linn.*	*	*	*	*	*			*					P. theseus type.
" helenus, *Linn.*	*	*		*	*	*		*					
" prexaspes, *Felder*.													
" iswara, *White*				*	*								
" saturnus, *Guérin*	*			*	*								
" Esperi, *Butler*				*	*								
" mestor, *Hübner*				*	*								
" achates, *Cramer*		*		*						*	*		
" varuna, *White*				*	*								
" diphilus, *Esper*	*			*	*					*	*		
" brama, *Guérin*							*	*	*				
" erebus, *Wallace*					*								
Casyapa thrax, *Linnæus*								*					Also from India,
" irava, *Moore*			*	*	*								Type from Sikkim.
Hesperia harisa, *Moore*													An allied species comes from Java.
" malayana, *Felder*							*						
" vitta, *Butler*					*								
" badra, *Moore*					*								Type from Bengal.
Cobalus olia, *Hewitson*						*							
" ciliatus, *Butler*.													

Species from Malacca.	Assam & Nepal.	Moulmein.	Ceylon.	Penang.	Singapore.	Borneo.	Sumatra.	Java.	Siau.	China.	South Pacific.	Australia.	Remarks.
Pamphila aria, *Moore*	..	..	*	..	..	..	..	..	..	..	..	..	Type from Bengal.
„ julianus, *Latreille*	..	..	*	..	..	*	*	*	..	..	..	..	
„ malthias, *Fabricius*	*	..	*	*	..	..	*	*	..	*?	..	..	
„ augias, *Linnæus*	*	..	*	*	..	..	..	..	*	..	..	*	
„ mœsoides, *Butler*	..	..	..	..	..	..	..	..	..	..	..	..	
„ maro, *Fabricius*	..	..	*	*	..	..	..	*	..	..	..	..	Also from S. India.
„ nigrolimbata, *Snellen*	..	..	..	..	..	..	..	..	..	..	..	..	
Plastingia callineura, *Felder*	..	..	..	..	*	*	..	*	..	..	..	..	
Cyclopides salsala, *Moore*	..	..	*	..	..	..	..	..	..	..	..	..	Type from Bengal.
Astictopterus jama, *Felder*	..	..	..	..	..	*	..	..	..	..	..	..	
„ xanites, *Butler*	..	..	..	..	..	..	*	..	..	..	..	..	
„ gemmifer, *Butler*	..	..	..	..	..	*	..	..	..	..	..	..	
„ armatus, *Druce*	..	..	..	..	..	*	..	..	..	..	..	..	
„ diocles, *Moore*	..	..	..	..	..	*	..	*	..	..	..	..	Type from Bengal.
„ sindu, *Felder*	..	..	..	..	..	*	..	..	..	..	..	..	
„ stellifer, *Butler*	..	..	..	..	..	..	..	..	..	..	..	..	
Plesioneura folus, *Cramer*	*	..	..	..	..	..	..	*	*	..	..	..	
„ alysos, *Moore*	..	*	*	*	..	..	*	*	..	..	..	..	
„ asmara, *Moore*	..	..	..	..	..	..	..	*	..	..	..	..	Also from the Khasia hills.
„ dan, *Fabricius*	..	..	..	*	..	..	..	*	..	..	..	..	Also from India.
„ Pinwilli, *Butler*	..	..	..	..	..	..	..	..	..	..	..	..	
Tagiades ravi, *Moore*	..	..	..	*	..	..	*	..	*	..	..	..	Type from Bengal.
„ gana, *Moore*	..	..	..	..	..	..	..	*	..	..	..	..	Type from Bengal.
„ calligana, *Butler*	..	..	..	..	..	..	..	..	..	..	..	..	
„ lavata, *Butler*	..	..	..	..	..	..	..	..	..	..	..	..	
„ trichoneura, *Felder*	..	..	..	..	..	..	..	*	..	..	..	..	Also from Bengal & Silhet.
258 species	65	38	33	04	46	112	41	87	39	26	2	0	

EXPLANATION OF PLATES.

PLATE LXVIII.

Fig. 1. *Thaumantis pseudoliris* (upper surface) ♂.
2. *Neptis peraka* (both surfaces).
3. *Neptis dorelia* (both surfaces).
4. *Plesioneura Pinwilli* (upper surface).
5. *Salpinx* (*Euplœa*) *leucoyonis* (upper surface) ♀.
6. *Neptis dindinga* (both surfaces).
7. *Papilio Esperi* (upper surface) ♀.
8. *Lampides pseudelpis* (upper surface) ♀.
9. „ „ (both surfaces) ♂.

PLATE LXIX.

Fig. 1. *Neptis gononata* (both surfaces).
2. *Neptis leuconata* (both surfaces).
3. *Neptis mamaja* (both surfaces).
4. *Athyma nivifera* (both surfaces) ♂.
5. *Athyma clerica* (both surfaces) ♂.
6. *Hestia linteata* (upper surface).
7. *Astictopterus xanites* (upper surface).
8. *Tagiades lavata* (upper surface).
9. *Euplœa Pinwilli* (upper surface) ♂.
10. *Zemeros albipunctata* (upper surface) ♂.
11. *Tagiades calligana* (upper surface).

Fig. 2.

S. rhia peridra

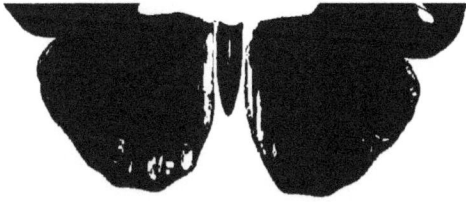

Drurmantes penteleus.

Fig. 3.

Fig. 4.

Planemos pusilla.

Fig. 5.

Saoma I. eucepia.

Fig. 6.

Neotis dinlemas

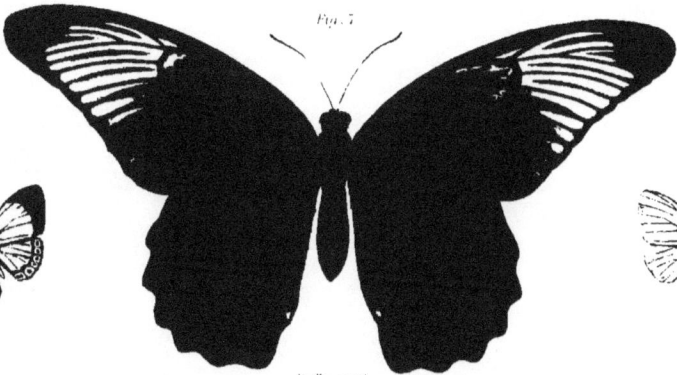

Fig. 7.

Fig. 8.

Fig. 9.

Aenpates pranceps.

Papilo esperi

BUTTERFLIES   MALACCA.

Fig. 1.

S. grisenata.

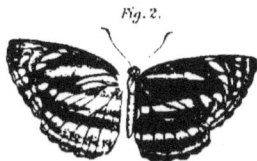

Fig. 2.

Neptis leucosata.

Fig. 3.

S. metanga

Fig. 4.

Athyma orelion

Fig. 5.

Athyma clearca

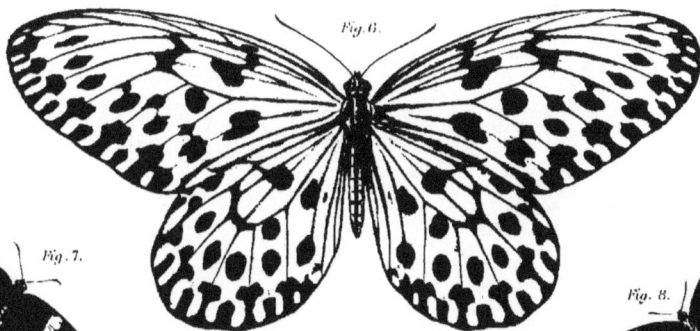

Fig. 6.

Hestia lineata.

Fig. 7.

Astictopterus xanites

Fig. 8.

Taglades lavata.

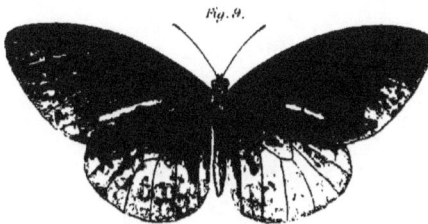

Fig. 9.

Euplœa pinmilla. ♂

Fig. 10.

Zemeros albipunctata

Fig. 11.

Zemeros allasona

www.ingramcontent.com/pod-product-compliance
Lightning Source LLC
Chambersburg PA
CBHW022032190326
41519CB00010B/1688